NAVAL
SHIPHANDLER'S GUIDE

TITLES IN THE SERIES

THE U.S. NAVAL INSTITUTE
BLUE & GOLD PROFESSIONAL LIBRARY

For more than 100 years, U.S. Navy professionals have counted on specialized books published by the Naval Institute Press to prepare them for their responsibilities as they advance in their careers and to serve as ready references and refreshers when needed. From the days of coal-fired battleships to the era of unmanned aerial vehicles and laser weaponry, such perennials as *The Bluejacket's Manual* and the *Watch Officer's Guide* have guided generations of sailors through the complex challenges of naval service. As these books are updated and new ones are added to the list, they will carry the distinctive mark of the Blue & Gold Professional Library series to remind and reassure their users that they have been prepared by naval professionals and they meet the exacting standards that sailors have long expected from the U.S. Naval Institute.

BLUE &GOLD
PROFESSIONAL LIBRARY

NAVAL
SHIPHANDLER'S GUIDE

Capt. James Alden Barber, USN (Ret.)

Naval Institute Press
291 Wood Road
Annapolis, Maryland 21402

ISBN 978-1-55750-435-7
Library of Congress Cataloging-in-Publication Data

Barber, James Alden, 1934–
  Naval shiphandler's guide / James Alden Barber.
      p. cm.
  Includes bibliographical references and index.
  ISBN 1-55750-435-0 (alk. paper)
  1. Ship handling. 2. Warships—Handling. I. Title.
  VK543.B24 2004
  623.88′1—dc22

                                                    2004009700

12 11 10   9 8 7 6 5 4
Fourth printing

CONTENTS

PREFACE

———◄○►———

Shiphandling, an ancient and honorable exercise of both art and science, has been one of the key skills of the professional naval officer for as long as there have been navies. It is a skill that must be learned, and it is to that process of learning that this guide is aimed. It is intended for the beginning and intermediate shiphandler. The old salt is unlikely to have need of the instruction contained within these covers, although it is hoped that there is nothing here to engender serious disagreement.

As the size of warships has increased and their numbers have dwindled, opportunities to handle ships have diminished. This is partly compensated by increasingly realistic shiphandling simulators. To get the most out of opportunities to practice shiphandling, whether on a simulator or a real ship, it is necessary to have done your homework. That homework is the purpose of this guide.

No one writes about shiphandling without owing an immense debt to those who have gone before. Knowledge of shiphandling has for centuries been passed from generation to generation, usually by apprenticeship and word of mouth but sometimes by the written word. My generation of shiphandlers was raised on the words of Capt. R. S. Crenshaw Jr. (*Naval Shiphandling*, 2nd ed. [Annapolis: U.S. Naval Institute, 1960]) and *Knight's Modern Seamanship*, as revised by Capt. John V. Noel Jr. (*Knight's Modern Seamanship*, 17th ed. [New York: Van Nostrand Reinhold, 1984]), and that debt is gratefully acknowledged. Over the years, the U.S. Naval Institute *Proceedings* has published articles and professional notes by serving officers who have taken the time and initiative to pass on for the good of the profession their shiphandling lessons learned. A number of these articles are listed in the bibliography. Notable among these authors is Vice Adm. James Stavridis, USN, who besides somehow finding the time to contribute revisions of *Watch Officer's Guide* and *Division Officer's Guide*, has made a cottage industry of passing on the shiphandling knowledge gained from his service on a variety of men of war. No single shiphandler is likely to have firsthand experience on every type of ship and in every circumstance. A debt is therefore owed and gratefully acknowledged to all of those earlier authors whose generous sharing of their knowledge has contributed to whatever merits this guide may have.

The author's largest debt is owed to Capt. Stu Landersman, USN (Ret.), whose unstinting assistance amounts almost to coauthorship. Besides multiple readings and invaluable suggestions as the manuscript took form, Captain Landersman contributed sections throughout, perhaps most notably the majority of the shiphandling vignettes that begin each chapter. Other major contributions were made by Rear Adm. W. J. Holland, USN (Ret.), who prepared the section of chapter 13 on submarines; Lt. Cdr. Joe DiRenzo III, USCG, who prepared the section of chapter 13 on Coast Guard cutters, and tug pilot Capt. Victor J. Schisler, who provided both advice on the use of tugs and pilots and the proposed tug standard commands and operating procedures to be found in Appendix B. Capt. Eric Shaw, USCG, was kind enough to review the Coast Guard section in chapter 13. Lt. Dennis Volpe, USN, provided a review of the entire manuscript from the junior officer's point of view.

The staff of Marine Safety International in both Norfolk and San Diego was of great assistance. In Norfolk, Capt. Brian Boyce, USN (Ret.), and Capt. Richard Williams, USN (Ret.), were particularly helpful. In San Diego, Rear Adm. David G. Ramsey, USN (Ret.), Capt. Donald F. Santamaria, USN (Ret.), Capt. Robert B. Lynch, USN (Ret.), Capt. Robert T. Glynn, USCG (Ret.), Capt. Robert L. Richardson, USN (Ret.), and Capt. Michael E. Mays, USN (Ret.), were unstinting in their assistance.

James R. Finney did a polished and professional job of preparing all of the line drawings to help clarify the text. Dawn Stitzel and Jennifer Till at the U.S. Naval Institute were invaluable in locating and choosing the right photographs to illustrate the text. Robert Allan Ltd. drew upon their expertise in commercial tug design to contribute the handsome drawings of tugboats to be found in chapter 8.

Thomas J. Cutler, Senior Acquisition Editor at the Naval Institute Press, conceived the idea of a new guide to naval shiphandling and gave me the encouragement to proceed. Judy Joyce provided both invaluable clerical assistance and help with an occasionally recalcitrant computer. Karin Kaufman provided highly professional copyediting.

The major portion of the credit for whatever is worthwhile in this guide must be given to the knowledge and generous assistance of the above-named experts. Blame for any and all errors of both omission and commission must go to the author.

NAVAL
SHIPHANDLER'S GUIDE

INTRODUCTION

—◁◇▷—

Their want of practice will make them unskillful, and the want of skill timid. Maritime skill, unlike skills of other kinds, is not to be cultivated by the way or at chance times.

Thucydides, 404 B.C.

For as long as there have been navies, shiphandling has been one of the most important professional skills. A well-handled ship excites the admiration of all observers and is a source of pride for the crew. No other aspect of command excellence is so visible, and few other professional activities are as personally satisfying to the practitioner.

As the number of ships in the Navy has shrunk and the size of the individual ship has grown, opportunities to practice the shiphandler's skills have diminished. There was a time when opportunities for a junior officer to handle his ship were frequent, and when a destroyer would be embarrassed to take a tug, let alone a pilot. But those destroyers displaced perhaps 3,500 tons and were relatively forgiving. A modern destroyer or cruiser has two to three times the displacement, more sail area, draws substantially more water, and has a large and vulnerable sonar dome projecting from the bow. Thus there has been a reduction in the occasions for unassisted shiphandling in constricted waters or in close proximity to other ships. To come to command with only limited experience in shiphandling is no longer as unusual as it once was. Fortunately this has been balanced, at least in part, by the development of excellent shiphandling simulators that permit realistic practice without hazard to the ship.

Shiphandling is both a science and an art. The science is in understanding the ship's physical characteristics and the forces that influence her movements.

Figure I–1. Shiphandling is one of the most important professional skills. *U.S. Navy photo by Chief Photographer's Mate Johnny R. Wilson*

Knowing how to use these characteristics and forces to move the ship into a desired position is an art. The shiphandler is an artist who employs scientific principles just as a painter uses variously colored and textured chemicals to create a portrait, a sculpture is shaped from stone using tools, and an operatic aria is formed of various audio frequencies. Just as artists combine scientific knowledge of chemistry, metallurgy, or sound with their own imagination, the shiphandler applies knowledge of technical scientific subjects to obtain a resultant objective such as the proper positioning of the ship, a product of his or her imagination.

Things You Need to Know

To become a capable mariner, the novice must first learn the physical characteristics of the ship: the length, beam, draft at different loads; type, number, power, and position of engines; type, number, and position of rudders; rpm and propeller pitch to speed; rudder angle to turn; surge distance; type, number, power, and position of auxiliary power/bow thruster; type, number, and position of mooring lines and anchors; and more. A novice must know these before he or she can expect to be given responsibility for maneuvering a ship. The experienced shiphandler faced with taking responsibility for an unfamiliar ship must learn these same physical characteristics so as to understand differences from more familiar classes of ships.

In addition to learning the physical characteristics of the ship, the novice seeking a higher level of achievement as a shiphandler must learn the academics: the technical scientific principles involved. The mariner must know physics, mechanics, hydrodynamics, meteorology, mathematics, and other academic sciences as they apply to shiphandling. He or she must read, study, and/or attend classes/seminars to gain an understanding of how to employ the various forces. Having previously learned the physical characteristics of a ship, the novice must now find out what happens when the propeller spins, the rudder goes over, or a mooring line takes a strain. Effects of wind and current must be understood. Physics includes studies of energy, power, work, and motion, with vectors and couples. How the propeller moves the ship through the water and why the rudder turns the ship are explained by principles of hydrodynamics, which must be understood by the shiphandler.

More Necessary Knowledge

The business of going to sea is, and always has been, subject to hazard. To reduce the risk of collisions and other accidents, detailed rules have evolved over the years. Many of these rules came from the accumulation of Admiralty law, many through agreements between states, and more recently through international organizations. The most important of these is the International Maritime Organization (IMO). The IMO, with headquarters in London, is a specialized agency of the United Nations dealing with maritime affairs. Since its organization after World War II, it has promoted the adoption of more than thirty conventions and protocols and adopted well over seven hundred codes and recommendations concerning maritime safety, the prevention of pollution, and related matters.

Many of the IMO regulations impact in one way or another on the operation of naval vessels, including the Convention on Safety of Life at Sea (SOLAS), the Convention on Maritime Search and Rescue, and the International Convention for the Prevention of Pollution from Ships (MARPOL). There are two things issued by the IMO with which the naval shiphandler needs to be familiar in detail. One is the International Regulations for Preventing Collisions at Sea (COLREGS), also known as the Rules of the Road. The second is the IMO Standard Marine Communication Phrases (IMO SMCP). The COLREGS are mandatory, need to be known in detail, and complied with to the letter. There are a number of good books providing the text of the COLREGS and helpful interpretation. The novice shiphandler should obtain one and make it the subject of careful study. An introduction to the Rules of the Road is provided in Appendix A, but no attempt is made in this volume to provide detailed information.

The IMO Standard Maritime Communication Phrases are advisory in nature, although under the International Convention on Standards of Training, Certification and Watchkeeping for Seafarers, as revised 1995, "the ability to use and understand the IMO SMCP is required for the certification of officers in charge of a navigational watch on ships of 500 gross tonnage or more."[1] For efficient communication of intentions and understandings with merchant ships, the shiphandler needs a good working familiarity with standard phraseology. Some discussion of this is set forth in chapter 7.

Seaman's Eye

As you go about learning the skills of a shiphandler you will hear references to "seaman's eye." Seaman's eye consists of the learned skills of timing and execution of planned maneuvers based on observation of all of the forces working on the ship. The skilled shiphandler plans his maneuver based on knowledge of his ship's characteristics, then measures by all means available to him, including seaman's eye, whether the maneuver is going according to plan. More than almost anything else it is this ability to detect and correct a deviation from plan that distinguishes the professional shiphandler.

The driver of a race car reacts instantly to sensory inputs. His time horizon extends only seconds ahead. A shiphandler needs to think much farther ahead. For any ship's evolution the sequence is the same: the conning officer must plan the maneuver, give the appropriate orders, observe whether the maneuver is proceeding as planned, correct deviations, and repeat the cycle until the maneuver is complete. To be able to do this requires an understanding of all of the forces working on the ship.

It has become customary to group the forces working on the ship into those that are under the control of the shiphandler and those that are not. Controllable forces include both those directly controllable from the bridge (engines, propellers, rudders, thrusters, and auxiliary power units) and those controlled with somewhat less precision by remote communication (tugs, lines, anchors). Forces not under the shiphandler's control, but which must be understood, and sometimes used, sometimes compensated for, include wind, current, and channel configuration. The capable mariner must know all of these forces and know how to use them. They are the tools used in the art of shiphandling.

To use these tools, the artist must learn a language, a professional naval/ maritime terminology that allows the shiphandler to control the forces as necessary to position the ship. As in mastering any language, the terminology must flow from thought to verbiage without going through discussion, translation or interpretation. The shiphandler must think shiphandling. He or she cannot think, I want the ship to move to the right. How do I make that happen? I think I should say something to position the rudder, like "right something rudder," but how much rudder should I use? Maybe "Standard" is correct, that's 15 degrees, I think, yes I'll say that. And then the shiphandler orders, "Right standard rudder." Too long. Not professional. The shiphandler must develop the capability to issue proper orders in response to thought. Just as the proper keys are struck by a computer operator to form a word without searching the keyboard or an automobile driver in traffic uses steering, brakes, clutch, and transmission without studying the position of each. Maneuvering and positioning of the ship must be accomplished by orders to the helm, lee helm, anchor detail, line handlers, and tugs using proper terminology, the language of shiphandling. The mariner must know how to give "Orders to the helm."

Once the mariner, novice or experienced, knows the physical characteristics of the ship, has learned the technical scientific principles involved, understands the many forces that affect maneuvering and positioning of the ship and has full use of proper terminology, he or she is ready to apply all that to shiphandling. But each individual has few opportunities to put all this into practice in a Navy ship. The novice will have to wait his or her turn. The commanding officer's support for an aggressive shiphandling training program for junior officers can be of great help, but chances to handle the ship still tend to be limited. In a typical two-year tour of duty, a Navy ship may make about forty landings alongside a pier. Pilots may be used twenty times, and the captain and executive officer might make five. That leaves fifteen landings for the remainder of the officers. Department heads may make six, leaving nine landings for the remaining twenty or so junior officers. The novice generally receives few opportunities and has little chance to achieve any measure of familiarity, competence, and confidence waiting for occasional, if any, actual shiphandling opportunities. Although no shiphandling training is as valuable as actual shiphandling, simulators offer the next most valuable opportunities. Use of the simulators is discussed in chapter 1.

Configuration of the Ship

The first step in learning to be a skilled shiphandler is to know the configuration of your own ship and how her behavior is affected by those characteristics. It makes a difference to the shiphandler whether his ship is single screw or multiple screws; whether single rudder or dual rudder, and whether the rudders are lined up directly behind the propellers or offset. When the ship is in drydock, climb down into the dock to get a first hand look at the configuration of your ship's screws and propellers. Gas turbine ships react differently than do steam turbine ships, and both differ from diesels. The capabilities and limitations of bow thrusters or auxiliary propulsion units are essential knowledge.

A ship with high sides and a large superstructure will be much more sensitive to the wind than one with less sail area. A ship with more sail area forward will react differently to the wind than one with more sail area aft. The shiphandler needs to know what anchors the ship has and the amount of chain for each anchor. To employ tugs you need to know whether there is a potential problem with sonar dome or screws. How you will handle your lines depends in part on the availability and location of capstans and their power. Knowledge of all of these characteristics of the ship and how they affect its behavior is a prerequisite for the beginning shiphandler.

Tactical Characteristics of the Ship

Every naval vessel has a folder of tactical characteristics, sometimes based on trials conducted with the individual ship, sometimes upon class characteristics. The data will include the ship's turning diameter for different rudder angles and different speeds, acceleration and deceleration tables, and so on. It is useful to commit at least some of the basic data to memory, and to carry more complete information in the pocket notebook or personal digital assistant no naval officer should be without.

Figure I–2. A clear mental image of the ship's underwater configuration is vital to the shiphandler. *Bath Industries*

Some tactical characteristics of the ship are best learned through experiment. An afternoon or two in independent ship's exercises at sea can do more to get a feel for how the ship holds her way while coasting, how she reacts to screws opposed at one-third and at two-thirds, how tightly she turns with various combinations of speed and rudder, how she lies to the wind, and so on than can any compilation of formal test results.

The Conn

One of the most important principles of shiphandling is that there should be no ambiguity as to who is controlling the movements of the ship. One person gives orders to the ship's engines, rudder, lines, and ground tackle. This person is said to have the "conn."[2] Transfer of the conn should be formal and needs to be communicated to everyone on the bridge, so that they know clearly for whose orders they should be listening. After receiving a turnover of all necessary information from the officer being relieved, the officer assuming the conn should salute and announce in a loud, clear voice to the bridge watch, "This is Mr. Smith. I have the conn." The helmsman and lee helmsman should acknowledge with, "Aye aye, Sir (or Ma'am)," followed by announcement of the gyro and compass courses being steered and the current order to the engines.

It will sometimes happen that the commanding officer or other senior officer authorized by the commanding officer issues an order without formally assuming the conn. The traditional understanding is that such an order constitutes an automatic assumption of the conn, but because of the possibility of confusion, it is the responsibility of the officer who had been conning to clarify the situation by asking, "Sir, do you have the conn?" Upon receiving the reply, then announce to the bridge either "The captain has the conn" or "This is Mr. Smith. I have the conn."

It is absolutely essential that the person with the conn not be distracted from controlling the ship. Watch responsibilities should be assigned such that other evolutions or administrative tasks do not place a claim on the conning officer's attention. Whenever possible the watch should be organized in such a way that a second qualified person is backing up the person with the conn, especially when the ship is maneuvering in formation or navigating restricted waters. For example, when the commanding officer has the conn, it is the responsibility of the officer of the deck (OOD) to assist with recommendations, watch for hazards, and guard against slips of the tongue.[3] Even the most experienced seaman is vulnerable to errors such as saying starboard when port is meant. An attentive backup has avoided many potential mistakes. When maneuvering it is also important to guard against the herd syndrome, in which everyone concentrates on the same thing. If the ship is turning to starboard, the conning officer should move to the starboard wing, but at the same time another competent member of the watch team should be on the port wing to avoid the possibility of a problem from the side unseen.

Few things are more important to professional shiphandling than a quiet bridge. The conning officer should never have to struggle to make him or herself heard above the noise of administrative tasks, private conversations, or other noise generators. The well-ordered bridge is characterized by quiet professionalism.

Shiphandling is one of the key measures of a naval officer. The captain of a navy ship is expected to be a capable mariner; skilled in tactics, navigation and seamanship; competent in logistics, administration, and employment of ship's armament; and expert in shiphandling. The captain must be capable of positioning the ship as necessary to carry out the mission that could be, as John Paul Jones put it, "in harm's way" or "alongside the enemy," or in more peaceful times, alongside a pier with adverse wind and current. Each shiphandling evolution must be accomplished by using the knowledge, judgment, and experience of the mariner. Knowledge can be gathered by books, lecture, seminar, and experience. Judgment may be honed through practice and experience. Experience is the key and each opportunity to gain shiphandling experience is a most valuable plank in the development of the naval officer. Each opportunity must be thought out and planned in advance, carried out with confidence, and afterward subjected to thorough critique and used as a learning tool.

There is great satisfaction to be taken in the skillful handling of a powerful and responsive warship. That skill comes only with investment of the necessary time and effort.

1

NAVAL SHIPHANDLING
SIMULATOR TRAINING

—◁○▷—

The local Navy scheduling authority, COMNAVSURFPAC, has scheduled USS *Higgins* (DDG 76) for a twenty-hour Shiphandling Availability at a shiphandling simulator facility in San Diego. Lt (jg) Don Cover, *Higgins*'s training officer, contacted the facility staff two weeks ago and learned that two simulators would be available, one a Full Mission Bridge Simulator, the other a Bridge Wing Simulator. Each simulator would be staffed by a facilitator/instructor and computer operator and would include a classroom. After discussing with the simulator facility staff the scope of services that would be available and appropriate for their ship, Don explained the arrangements at a meeting of the *Higgins*'s Training Board and prepared a complete training plan for their Shiphandling Availability. The plan as approved by *Higgins* includes use of both simulators for the full twenty hours, identifies personnel who will participate, and gives preparatory training and homework.

The *Higgins* training plan directs the facility staff to set up the Full Mission Bridge Simulator with a scenario for getting under way in daylight from the north side of Pier 7, San Diego Naval Station, starboard side to, bow in, with two cruisers astern, 6 knots of wind from 340 and .2 knots of flood current. One tug will be available to assist. Harbor traffic is moderate. *Higgins* will transit San Diego Bay to sea during sunset so that final legs passing the submarine base and Ballast Point will be in darkness. The plan includes reversal of the under way and transit, in which the ship will enter port, proceed to the naval base, and moor at the same pier. *Higgins* will provide radar navigation team, helm and lee helm, and the ship's navigator will pre-brief the exercise, observe the process, and conduct critique in the classroom with monitor and printed replay.

The Bridge Wing Simulator will be used for conducting daylight underway replenishment (UNREP) exercises in a moderate sea state. *Higgins* will be at sea within three miles of a formation of ships that includes the replenishment ship *Yukon*. *Higgins* will approach on a reciprocal course, will maneuver to take position astern of *Yukon*, approach at proper speed and lateral separation, take and maintain position alongside, and clear the side when fueling is completed. Proper signals will be used. *Higgins* will be prepared to respond to propulsion and/or steering casualties. The shiphandling simulator facility staff will provide all off-bridge wing functions and a facilitator/instructor will pre-brief the exercise and conduct a classroom critique with monitor and printed replay.

Personnel will be rotated through different positions and between simulators as the exercises are repeated. In the final four hours of the Shiphandling Availability, the Full Mission Bridge will be used for high-density shipping management, and the Bridge Wing Simulator will be used for junior officer pier landings.

Prior to the scheduled Shiphandling Availability, *Higgins* will conduct training sessions on board ship covering maneuvering characteristics of the ship, proper orders to line-handling stations, helm and lee helm, and the use of signal flags for underway replenishment. Individuals who will participate in the training have been given reading assignments.

At the day and time of the scheduled Shiphandling Availability, personnel of *Higgins*, including the captain, will arrive at the shiphandling facility. They made contact previously with the facility staff, learned what was available, promulgated a full training plan, and made it available in advance to all involved, including the simulator staff. Now they have done their homework and are ready to start training. The *Higgins* plan includes exercises in getting under way, mooring to a pier, harbor egress and entry, and underway replenishment, making excellent use of the simulators.

The most important and valuable advancement in shiphandling training has been the introduction of shiphandling simulators. More than fifty simulators that meet international standards exist in the United States, and there are more than two hundred worldwide. In addition to these full-sized devices, there are countless simulators of lesser capability, as well as model boat basins, that offer valuable training opportunities for shiphandlers to learn and to improve proficiency. For training in shiphandling at fleet concentration areas, the U.S. Navy has available high-quality, realistic shiphandling simulators that meet international standards. In addition, the U.S. Navy's Navigation, Seamanship and Shiphandling (NSS) Trainer Project includes the shipboard and classroom laptop or PC-based Bridge Training System, which will give each ship an on-board shiphandling training capability. All simulators of the various types have value, and as the Full Mission Bridge Simulator that meets international standards is the most valuable, it will be discussed here as applicable to naval shiphandling.

Shiphandling Simulator Facilities

Navy ships receive the benefits of shiphandling simulation training services from shiphandling simulators in the major fleet concentration areas. The training provided by these simulators is essential for the safe operation of Navy ships. Ships' personnel receive this training to learn, maintain, and improve proficiency in shiphandling skills, to sharpen their safe navigation procedures, and to reduce the probability of collisions and groundings. Without this training, Navy ships operate at higher risk. In addition to providing the highest quality of shiphandling simulation services, in response to Navy requests the simulator operators have developed specialized training courses in safe navigation and collision/grounding avoidance. Courses are offered in use of the latest electronic charts and automatic radar plotting aids, and the integration of these systems, and in Bridge Resource Management (BRM). These specialized training courses meet International Maritime Organization standards, are based on and certified by the U.S. Coast Guard, modified and certified by the Navy, and are of great value to Navy shiphandlers.

Each shiphandling simulator complex includes at least a simulator, classroom, and learning laboratory. The shiphandling simulator is a full-sized stand-up and walk-around replica of a Navy ship's bridge, pilot house, and/or bridge wing with projected on screen large-scale video display providing realistic, animated, and responsive seascape. Classrooms equipped with playback are used for critique of shiphandling evolutions, seminars, and courses of instruction related to shiphandling, safe navigation, and collision/grounding avoidance. Instructions in setup, application, integration, and uses of systems such as Electronic Chart Display and Information System (ECDIS) and Automatic Radar Plotting Aids (ARPA) are provided in the Learning Laboratory.

Figure 1–1. The Full Mission Bridge Simulator is an invaluable training aid for the shiphandling student. *Maritime Safety International*

Training Services

The full programs of shiphandling simulator proficiency training services, training in safe navigation, and training in measures to reduce the probability of collisions and groundings are available at the shiphandling simulator complexes to personnel serving in Navy ships. This training includes instructors/facilitators with experience and qualifications certified in accordance with International Maritime Organization and U.S. Coast Guard standards. Both certification standards require that an instructor must hold at least the professional credential required of the student. At the shiphandling simulator facilities, these qualifications have been applied to Navy training so that in a class that includes a Navy ship's commanding officer, the instructor has had that and often more than that level of professional experience. Most of the instructors have had multiple sea commands in Navy, Coast Guard, or merchant ships. In addition to experiences in command at sea and U.S. Coast Guard and U.S. Navy qualification/ certification, these instructors have experience in teaching shiphandling, safe ship navigation, and seamanship in a training program that includes shiphandling simulation and classroom and laboratory work.

Available Simulations

The shiphandling simulators at Navy fleet concentration areas are capable of simulating total shipboard bridge operational situations, including open ocean maneuvering and advanced maneuvering in restricted waterways, for safe navigation and collision avoidance. The simulators are capable of simulating a realistic environment for the development and maintenance of proficiency in various shiphandling skills:

1. Planning and conducting a passage and determining position;
2. Maintaining a safe navigation watch;
3. Using radar to assist in safety of navigation;
4. Responding to emergencies;
5. Maneuvering and handling the ship in various conditions;
6. Establishing watch arrangements and procedures.

The shiphandling simulators include equipment and consoles installed, mounted, and arranged in a Navy shiplike manner. The following equipment is included in the simulators:

1. Controls of propulsion plant operations, including engine order telegraph, pitch control, and thrusters (indicators for shaft revolutions and propeller pitch and controls for propellers and thrusters);
2. Controls of propulsion plant for mooring operations;
3. Steering console resembling that of a Navy ship with indicators of rudder angle;
4. Steering compass and bearing compass with an accuracy of at least 1 degree;
5. Radar and ARPA;
6. Communications equipment including intraship and VHF radio;
7. GPS, echo-sounder, and speed log;
8. Instruments for indication of relative wind direction and force;
9. Sound panel for ship's whistle signals.

The shiphandling simulators provide a high degree of behavioral realism through accurate computer ship models for each Navy ship class:

1. Accurate models are provided for every Navy ship class as own ship.
2. The simulation of own ship is based on a mathematical model with 6 degrees freedom of movement.
3. The models realistically simulate own ship hydrodynamics in open water conditions, including the effect of wind forces, wave forces, tide, and current.

4. The models realistically simulate own ship hydrodynamics in restricted waterways, including shallow water and bank effects, and interaction with other ships.

5. The radar simulation is capable of modeling weather, tidal current, shadow sectors, spurious echoes, and other propagation effects. The radar generates traffic ships, coastlines, and navigational buoys.

6. The ARPA simulation includes
   a. Manual and automatic target acquisition;
   b. Past track information;
   c. Use of exclusion areas;
   d. Vector/graphic time-scale and data display;
   e. Trial maneuvers.

7. The simulators provide own ship engine sound, reflecting the power output.

The shiphandling simulators provide a realistic operating environment through accurately modeled data bases of real-world ports and operating scenarios:

1. Data bases for most of the real-world harbors and ports frequented by the Navy are provided.

2. At least twenty different types of traffic ships are provided, each based on a mathematical model that provides motion, drift, and steering angles according to forces induced by current, wind, and wave.

3. Traffic ships are equipped with appropriate navigational lights, shapes, and sound signals.

4. The simulators provide a realistic visual scenario by day, dusk or night including variable meteorological visibility, changing in time. It is possible to create a range of visual conditions, from dense fog to clear.

5. The visual system replicates movements of own ship according to 6 degrees freedom of movement.

6. The visual scene has an update rate such that the scene is smooth with no jerkiness.

7. The visual scene displays are at such a distance and viewed in such a manner from the pilot house or conning area that accurate visual bearings may be taken to objects in the scene.

8. The visual system presents the outside world by a view of at least 220 degrees in the horizontal and the rest of the horizontal view can be panned through 360 degrees.

9. The visual system presents at least 30 degrees vertical field of view.

10. The visual system presents navigational markings and aids according to charts used.

11. The visual system shows objects with sufficient realism, detailed enough to be recognized as in real life.

12. The simulators provide environmental sounds according to conditions simulated.

13. The navigated waters include currents according to charts used and reflect tidal effects.

14. The simulation includes depth according to charts used, reflecting water level according to tidal conditions.

15. The simulators provide realistic sea states with surface waves variable in strength.

Simulator Benefits

Shiphandling simulators provide Navy ship personnel with

1. Multiple opportunities for basic shiphandling proficiency development;

2. Approved, certified, and practiced specialized training courses in safe navigation and collision and grounding avoidance;

3. Accurate and validated databases of the most frequented U.S. and overseas ports under all environmental conditions;

4. Accurate and validated hydrodynamic ship response models of every Navy ship class;

5. At-sea exercise scenarios including the most often used Navy formations, traffic ship management, and multiple ship operations under all environmental conditions;

6. Highest level of qualified and experienced professional facilitators/instructors;

7. Leading edge of technology shiphandling simulator systems and full specialized training programs.

Preparation for a Shiphandling Availability

Each shiphandling simulator facility is overseen and is scheduled by a Navy command. The Navy command schedules students and/or ships for use of the shiphandling simulator facility. Typically, in schoolhouse situations such as Surface Warfare Officers' School in Newport, Rhode Island, the school sends students to the shiphandling simulator facility for pre-planned shiphandling training. At major fleet concentration areas such as San Diego, California, and Norfolk, Virginia, a local Navy command represents the fleet commander in scheduling ships for use of the shiphandling simulator facility. The time a ship is scheduled for the facility is called a Shiphandling Availability, and each ship so scheduled determines how the facility will be used.

It is important that each student take full advantage of the training opportunity. Each ship scheduled for a Shiphandling Availability needs to fully utilize their time at the shiphandling facility, getting the most benefit from simulators, classrooms, and instructors. A missed class or unused time is a waste of a valuable resource. If your ship is scheduled for twenty hours of shiphandling simulator use, make sure that you use all twenty hours. Obtaining the most value from a shiphandling simulator requires planning and preparation before you get there and application once there.

Simulator facility use gives a Navy ship the rare opportunity to conduct training in basic shiphandling, safe navigation, and collision/grounding avoidance. The ship should prepare in advance of the scheduled time, so that the maximum benefit can be realized. Preparation for scheduled use of the shiphandling facility should include the following:

1. Making contact early with the shiphandling simulator staff. Early contact confirms the schedule and allows the facility staff to be ready with equipment set up for proper scenarios and with prepared classes. Make early contact so that both ship and training facility are prepared.

2. Discussing what services are available. It is important that the ship scheduled for use of the facility learn what is available. Often there will be more than one simulator available, with additional classrooms and instructors. Simulator facilities can provide specialized/certified classroom instruction in various courses related to safe navigation and collision/grounding avoidance, in addition to the basic shiphandling proficiency opportunities. Find out what the shiphandling simulator facility has to offer.

3. Determining a training program that utilizes the full resources of the facility. Once the ship has been scheduled for use of the shiphandling facility, and has learned what is available, it is important that the ship lay out a complete training plan. The plan must utilize the full resources available at the facility, for the complete scheduled time, and accomplish the intended training objectives. It should identify the ship class and port to be used, exercise scenarios, environmental conditions, traffic and personnel that will participate. This training plan must be made available to the shiphandling

simulator facility staff prior to the scheduled training. Make a complete training plan. A sample training plan is provided below.

4. Completing appropriate "homework." Any preparatory training that can be accomplished on board ship or away from the shiphandling facility should be completed before the scheduled time in the facility. Instruction in every facet of shiphandling, safe navigation and collision/grounding avoidance can be provided by the training facility, but many of the training requirements can be accomplished elsewhere. If these fundamentals are done elsewhere, the full value of the simulator facility can be directed at those areas that the facility does best. Time taken to teach fundamentals such as own ship's characteristics and "orders to the helm" can be better spent practicing shiphandling. The facility will provide whatever training the ship requests, but the ship must determine if their time could be put to better advantage if "homework" is accomplished prior to the scheduled shiphandling simulator facility visit. Some of the topics that might be included as "homework" and covered as review are ship's characteristics, Rules of the Road, hydrodynamics, captain's standing orders, orders to the helm/lee helm, mooring lines, anchors, tugs, electronic charts, and radar/ARPA.

5. Showing up at the facility on time with the proper personnel. When it's time to commence scheduled training the ship should ensure that the proper personnel are at the shiphandling facility, and for the full scheduled time. The training plan should identify personnel who will conn the ship, as well as navigate, handle the helm and lee helm, instruct, and observe. In using shiphandling simulation services "time is money" and failure to utilize the scheduled time completely and with appropriate personnel is a waste of money. Be there on time, for the full time, and with the right people.

Using the Shiphandling Simulator Facility

Much of the value you obtain from use of the shiphandling simulator depends on the advance preparation of a good training plan, including the selection of appropriate scenarios. Exercise scenarios most often used in shiphandling simulators are

1. Getting the ship under way from alongside a pier with considerations of wind, current, water depth, pier congestion, traffic, and visibility, with and without tugboat assistance;

2. Mooring the ship to a pier with the same considerations;

3. Harbor/port entry and egress with the same considerations and including transiting a channel, piloting and visual navigation, and Rules of the Road;

4. Underway replenishment, including approach, positioning, lateral separation, and signals;

5. Traffic management, including maneuvering in high-density shipping traffic, Rules of the Road, and traffic separation schemes;

6. Anchoring and mooring to a buoy;

7. Specialized procedures, including operating in formations and with aircraft carriers and emergency shiphandling.

These exercise scenarios can be provided with a full spectrum of environmental conditions, such as varying daylight from bright to dark and any sea state, wind, and current. Every Navy ship class is an accurate hydrodynamic model of your ship, with Bernoulli and Venturi effects, bottom and sidewall effects, turning and stopping, as your ship does. Even with the fine realism of Full Mission Bridge shiphandling simulators, nothing is as good, as realistic, as handling your own real ship. Simulators come close, and offer more opportunities, and once you have mastered the handling of your ship in a simulator, you are ready to handle a real ship. The following is a sample of a ship's simulator training plan:

Higgins is scheduled for a twenty-hour Shiphandling Availability 8–10 December at the COMNAVSURFPAC Shiphandling Facility, Building 1349, Naval Station San Diego. The Facility will have two simulators available, Full Mission Bridge (FMB) and Bridge Wing Simulator (BWS). The ship will provide the instructor for some exercises, and the captain will participate. The facility staff will provide a facilitator/instructor as requested by the ship. The facility will be used as follows:

Monday, 8 December
0800–1200

FMB	Team A	Under way from pier and harbor egress	Navigator instructing
BWS	Team B	Underway replenishment	Facilitator instructing

1300–1700

FMB	Team B	Under way from pier and harbor egress	Navigator instructing
BWS	Team A	Underway replenishment	Facilitator instructing

Tuesday, 9 December
0800–1200

| FMB | Team A | Harbor entry and pier landing | Navigator instructing |
| BWS | Team B | Underway replenishment | Facilitator instructing |

1300–1700

| FMB | Team B | Harbor entry and pier landing | Navigator instructing |
| BWS | Team A | Underway replenishment | Facilitator instructing |

Wednesday, 10 December
0800–1000

| FMB | Team A | Traffic management | Facilitator instructing |
| BWS | Team B | Pier landing | Facilitator instructing |

1000–1200

| FMB | Team B | Traffic management | Facilitator instructing |
| BWS | Team A | Pier landing | Facilitator instructing |

Figure 1–2. The Full Mission Bridge Simulator provides a realistic environment for the shiphandler. *Maritime Safety International*

Teams

Team A: Lieutenant (jg) Custodi, Lieutenant (jg) Waggoner, Lieutenant (jg) O'Connor, Ensign Hibler, Ensign Hutcheson
Team B: Lieutenant (jg) Pierpont, Lieutenant (jg) Desrocher, Ensign Sexton, Ensign Allen, Ensign Whitenight

Exercises

Under way from pier and harbor egress:
Ship will get under way in late afternoon daylight from San Diego, starboard side to Pier 7 North, bow in, with 6 knots of wind from 340, .2 knot of flood current toward 140, increasing to .5 knot in the channel, one tug assisting. Two cruisers nested astern. Two amphibs across the slip south side of Pier 6. Ship will transit harbor to sea, darkening, moderate traffic.

Harbor entry and pier landing:
Ship will transit in daylight from sea into San Diego and moor starboard side to Pier 7 North, bow in, with 6 knots of wind from 340, .5 knot of flood current in the channel decreasing to .2 knot toward 140 in the slip, one tug assisting. Two cruisers astern. Two amphibs across the slip south side of Pier 6.

Underway replenishment:

Exercise will commence with ship at sea on course 185 speed, thirteen knots en route to join formation for refueling with oiler *Yukon*. *Yukon* bears 230, 4,500 yards, and is on Romeo Corpen 005, speed thirteen knots. Ship will maneuver to take waiting station, then make approach on *Yukon* and take proper fore and aft position alongside with 150-foot lateral separation. Appropriate signal flags will be used. When fueling is completed, ship will clear *Yukon* side and proceed to next assignment.

Traffic management:

Ship will be making transit of waterway with high density merchant ship traffic. Numerous situations will require knowledge of the Rules of the Road. Proper judgment must be exercised in safely maneuvering the ship through difficult situations. Proper communication procedures will be observed in communicating with merchant ships.

Pier landing:

Exercise for the most junior officers. Ship will moor starboard side to Pier 7 North, bow in, with 6 knots of wind from 340, .5 knot of flood current in the channel decreasing to .2 knot toward 140 in the slip, one tug assisting.

On-board ship training:

Wednesday, 3 December 1300–1400. All personnel of Teams A and B assemble in the pilot house with navigator for training in orders to the helm and lee helm and UNREP signals. Group will then go to fo'c'sle with first lieutenant for training in line-handling commands.

Homework:

All personnel of Teams A and B are to reread *Watch Officer's Guide*, chapters 5, 6, 7, and 8, Ship's Characteristics, and Rules of the Road (COLREGS) prior to 8 December.

Guidance:

Be on time. Stay alert. Pay attention. Use all of your time wisely in the simulator. This is a great opportunity to learn, practice, develop, and improve your shiphandling skills. The simulator is not as realistic as the real ship, but almost!

Specialized Training at the Simulator Facility

Specialized training courses provide instruction in safe navigation and collision/grounding avoidance. These Navy-approved specialized training courses are taught in accordance with government standards and, in most case, international standards by certified, qualified, and experienced instructors in conjunction with approved curricula.

Specialized training in safe navigation and in procedures to reduce the probability of collisions and groundings are provided at shiphandling simulator facilities through a program of training courses. This program conveys to the shiphandler student the use of all resources and navigation capabilities of the ship.

The specialized training program includes at least three courses of particular interest to the naval shiphandler:

1. Bridge Resource Management Course. Derived from international and U.S. Coast Guard–approved courses, this course is modified for Navy applicability. Analyzes Navy maritime collisions and groundings, reviews teamwork considerations, and factors in sound decision making. Provides skills necessary to increase situational awareness, recognize development of and be able to break the error chain, develop sound voyage plans, and understand effects of stress and fatigue on watch standing.

2. Electronic Chart Display and Information System Course. Derived from international and U.S. Coast Guard–approved courses, this course is modified for Navy applicability. Covers principal ECDIS systems: charts, GPS, gyro, log, radar, ARPA, and fathometer. Teaches the naval shiphandling student

    a. Installing and starting ECDIS;
    b. Operational use of electronic charts: scale, day/night, layers;
    c. Route planning;
    d. Route monitoring: cross track error, waypoints, danger areas, alarms;
    e. Chart updates;
    f. ECDIS with ARPA.

3. Automatic Radar Plotting Aids Course. Derived from international and U.S. Coast Guard–approved courses, this course is modified for Navy applicability. Covers principal ARPA systems: displays, vectors, and readouts. Teaches the naval shiphandling student

    a. Plotting techniques;
    b. Target acquisition;
    c. Capabilities and limitations;
    d. Displays and target information;
    e. Errors of interpretation of displayed data;
    f. Performance checks against manual plots;
    g. Trial maneuvers;
    h. Navigation functions.

Bridge Resource Management

A Navy ship operating in restricted waters normally is at "special sea and anchor detail," with about twenty qualified people manning all ship control and navigation stations in the traditional legacy bridge/pilot house area, each with a carefully defined responsibility, each trained and ready to perform his/her duties. The Integrated Bridge System (IBS) could have fewer, perhaps only captain and conning officer, but for the ship to operate safely, to avoid a collision or grounding, all of the ship control and navigation functions performed in the traditional legacy bridge/pilot house must be accomplished in the IBS. Twenty or two, each participant in ship control and navigation must understand his/her duties and responsibilities, each must be trained in individual functions, and, just as important, all must function as a team.

Bridge Resource Management is a team-training program for key ship control and navigation personnel. The objective of Navy Bridge Resource Management is to provide ships' personnel with the skills necessary to avoid collisions or groundings.

The BRM program shows ship personnel how to increase situational awareness, recognize the development of and be able to break the error chain, develop sound voyage plans, and understand the effects of stress and fatigue on watch standing. Participants review communications skills, review teamwork and decision making, perform shiphandling, and develop an understanding of the captain-pilot relationship.

The aims of the prudent mariner are to ensure that the ship reaches its destination safely and that seamanship evolutions are conducted efficiently. To do this consistently demands a level of skill that must be part of the Navy culture. An accident by its nature is unexpected, but most collisions and groundings occur because the system in operation to detect and to prevent mistakes is dysfunctional. Studies show that more than 80 percent of maritime property damage claims and more than 90 percent of collisions are directly attributable to personnel error. In the past ten years, almost every Navy collision and grounding has been the result of personnel error. Collisions and groundings are rarely the result of a single event. They are invariably the result of a series of nonserious incidents, the culmination of an error chain.

To achieve a sound and efficient bridge/pilot house organization, procedures must be established to minimize risks. Most collisions and groundings could have been avoided with a properly functioning team approach. BRM is more than a concept, it is the implementation of a way of working that recognizes that reliable and consistent standards can best be maintained if watchkeeping is based upon sound principles and reinforced by effective organization. All ships' watch-standers must make the best possible use of available resources, both human and material.

At sea, on the bridge of a ship, watch-standers have to work together and help the conning officer make proper decisions, and developing teamwork among bridge watch-standers requires both training and practice. BRM encompasses classroom sessions, simulator exercises, and case studies of Navy collisions and groundings. A ship's bridge/pilot house personnel, including the commanding officer and other senior officers, are the attendees.

As training time at sea is reduced, the burden of training watch-standers increases and individual skills and teamwork must be developed and maintained. This is especially true as bridge teams are reduced as a result of "smart ship" technology and with the introduction of IBS. The internationally accepted program of BRM tailored for Navy team training in safe navigation, seamanship, and shiphandling is one of the most valuable tools to help the ship avoid collisions and groundings.

Bridge to the Future

As new ships with new weapon and propulsion systems are joining the fleet in an age of increased technology, modern methods are being introduced to utilize the latest technology for greater reliability and efficiency with fewer personnel. New concepts are being introduced that are changing bridge/pilot house configuration, navigation methods, and shiphandling control systems. Even the well-established method of giving "orders to the helm" is no exception, and the time-honored process of oral/verbal order, acknowledgment, and accomplishment may be altered with new technology.

The newest Navy ship can be expected to be equipped with an Integrated Bridge System. As defined by Navy General Specifications for Shipbuilding, Section 438, the Integrated Bridge System consists of

1. Voyage Management System—a computer based navigation, planning, and monitoring system;
2. Automated Radar Plotting Aid—a system that automatically acquires and tracks contacts;
3. Ship Control System—a steering and thrust control module.

The Integrated Bridge System is a combination of interconnected systems that allow centralized access to sensor information and command/control systems from a workstation, with the aim of increasing safe and efficient ship's management by qualified personnel. Navigation and ship control functions have likewise been centralized to make maximum use of automation and electronic capabilities. Navigation capabilities have been enhanced by the installation of a Voyage Management System (VMS) that uses an ECDIS as a core, taking inputs from the global positioning system (GPS), ship's gyrocompass, ship's pitlog, fathometer, and radar including ARPA. In the IBS, the VMS can drive an autopilot along a pre-programmed route with

Figure 1–3. The bridge of the future may have more in common with an aircraft cockpit than with the traditional warship's bridge. *U.S. Navy photo*

pre-developed parameters. The IBS has sensors in critical internal and external components and the ability to monitor those sensors networked to the bridge. In addition to monitoring the sensors, the watch officer can take steps to stabilize abnormal conditions or alert response teams, as necessary.

In IBS, the watch-stander is the conning officer and the officer of the deck, the direct representative of the commanding officer, and he or she is faced with several differences in the way that these watches were stood in traditional legacy bridge configurations. The first and most significant difference is that a man-machine interface now becomes the focus of decision making. Not only must the watch-stander have traditional proficiency in maritime skills, but he/she must now be proficient in all aspects of multi-computer-generated data. A second major difference between a legacy bridge and an IBS bridge is that the IBS watch-stander must seek out the electronic data, analyze it, and then act on it, often without any other human input.

A third difference is that the IBS watch-stander, using an automated VMS, makes decisions and controls the ship by exception rather than by traditional verbal orders and responses. A new mindset is required of the IBS watch-stander.

Future IBS physical configurations may resemble an aircraft cockpit more than the traditional legacy ship's bridge/pilothouse. In a future configuration, the watch officer may be able to roam from the centralized workstation with a remote controller in hand, or the controls may utilize and respond to voice recognition technology.

The IBS demonstrates significant potential for payoff in terms of improved underway voyage management and shiphandling, with a reduction in personnel workload and numbers. More than ever before, the watch-stander must be fully qualified before he/she assumes the watch, ready for increased responsibilities.

Electronic Chart Display Information System

The future of paper charts in the U.S. Navy is limited. Soon there will be no more taking bearings, drawing lines, analyzing tiny triangles, and looking at where the ship has been. Soon electronic charts with GPS will be used and the shiphandler will see where he/she is and where he/she is going.

The core of the Integrated Bridge system is the Electronic Chart Display Information System, an electronic chart system that meets IMO standards. There are many "off-the-shelf" electronic chart systems available for purchase, installation, and use by ships, and most would have value to Navy ships, but it is important to understand the difference between any electronic chart system and ECDIS.

There are two kinds of electronic charts, raster and vector. A raster chart is a digitized picture of a paper chart in which all data are displayed in one layer and one format, usually limited to four colors and displaying all data continuously. A vector chart has all data organized into separate files/layers, can display 256 colors, and displays only selected data. The IMO recognizes only vector chart format as meeting the standards for certification, so all ECDIS are electronic chart systems but not all electronic chart systems are ECDIS.

Raster charts are cheaper, data constrained, zoom limited, permanently cluttered, and they cannot be queried. Vector charts are smarter. They can be repositioned, their data can be tailored, queried and decluttered, and they are unconstrained as to zoom.

Like all computer-based systems, the accuracy of ECDIS depends on the quality of the information it receives and the quality of its installed databases. Installed vector charts are the best quality worldwide charts, usually including not only U.S. data but also those from the British Admiralty, Canada, France, Sweden, and Russia. Accurate ship's position input comes through GPS and maneuvering information is provided by ship's gyrocompass, pitlog, fathometer, and radar.

The U.S. Navy is looking at and testing a number of electronic chart systems, including some ECDIS systems, and expects to settle on one standardized version. Until one system is selected for all ships, Navy shiphandling personnel will have to be trained and become expert in the particular system found in their ship. It can be expected that ultimately Navy ships will be equipped with ECDIS meeting IMO performance standards. Those standards are the following:

1. Primary function of ECDIS is to contribute to safe navigation;
2. Capable of displaying all chart information necessary for safe and efficient navigation;
3. Reduce the navigation workload as compared to use of a paper chart;
4. Enable user to execute in a timely manner all route planning, monitoring, and positioning;
5. Capable of continuously plotting ship's position;
6. Have same reliability and availability of presentation as authorized paper chart;
7. Provide appropriate alarms regarding displayed information or equipment malfunction;

8. Capable of route planning on one chart while simultaneously monitoring navigation mode.

For maximum value in the IBS, ECDIS should be connected in an integrated conning system for collision and grounding avoidance. In full integration, the ECDIS display can be configured to show precise ship position and any and all navigation data on any chart, any scale, in the worldwide database, along with route planning information. With ARPA integrated, the full surface traffic picture can be included on the ECDIS display.

Automatic Radar Plotting Aid

Many Navy ships are being fitted with Automatic Radar Plotting Aid radar systems, and ARPA is included in the Integrated Bridge System. ARPA is a system that extracts data from raw radar information and presents an analysis of the radar data. It automates the operation of plotting traditionally carried out on paper or on a plotting screen or repeater head by lead or grease pencil. ARPA reduces collision potential by providing the conning officer with a great deal of surface-ship track information, performing most of the traffic management advisory functions provided by the Surface Section of the Combat Information Center (CIC).

ARPA maintains surveillance of traffic ships in track by a system processor, and allows the operator to try avoidance maneuvers before actual execution while continuing surveillance of traffic in track. While ARPA helps the conning officer with plotting and risk analysis, it has no Rules of the Road input, and it takes a professional mariner with traditional knowledge, experience, and skill to interpret and apply appropriate rules.

The International Maritime Organization and the U.S. Maritime Administration (MARAD) require ARPA in merchant ships of 10,000 or more gross tons, with detailed performance standards, including detection and presentation quality, tracking of numerous targets, and traffic ship information. Navy ships are not required to meet IMO or MARAD standards, but those equipped with ARPA can expect the same performance. ARPA performance standards are the following:

1. Manual or automatic target acquisition with clear display indication;
2. Track, process, display, and update at least twenty targets simultaneously;
3. Selection criteria for guard and exclusion zones;
4. Display of all data provided by a standard radar display;

5. Controlled range scales;

6. Capable of relative motion display with either north-up or course-up;

7. Course and speed vectors indicating target predicted motion, relative or true, adjustable;

8. Adjustable brilliance of screen;

9. Display clearly visible to more than one observer, day or night;

10. Range and bearing to any target;

11. Predict target motion within one minute, high accuracy within three minutes;

12. Range, bearing, closest point of approach (CPA), time of closest point of approach (TCPA), true course, and speed of any target;

13. Warnings for untracked, dangerous, and lost targets, and for wrong request;

14. Capable of simulating effect on all targets of a proposed maneuver;

15. Rapid recalculation of all target information upon completion of maneuver;

16. Installed test programs;

17. Connectivity as an integrated conning system for collision avoidance;

18. Parallel indexing for safe navigation;

19. Sea and ground stabilization.

2

FORCES ON THE SHIP

————◄○►————

USS *Mason* (DDG 87) had just secured from special sea and anchor detail after tying up alongside the pier in its home port of Norfolk, and the ship's officers had assembled in the wardroom. To get the most training benefit from each shiphandling opportunity it was the captain's policy to have a brief hot wash-up meeting after each significant evolution, chaired by the officer who had the conn. Today's meeting was chaired by Lt (jg) Roger Kingsbury.

"Let's get started," said the commanding officer, Cdr. Welsh Jones. "Roger, how did you think it went?"

"I screwed it up skipper," replied a somewhat embarrassed Lieutenant (jg) Kingsbury.

"Don't worry about it," said Commander Jones. "Nothing got bent, and we always learn more from our mistakes than from things that go perfectly. What do you think went wrong today?"

"I think I understand it now. I failed to properly anticipate the effect the current would have on us. I knew that with a strong ebb tide we would have a current setting us to port. Since we were going starboard side to, I planned to make a close approach to compensate for the current. What I didn't anticipate was that as we entered the slip the current would no longer work on the bow but would continue to push our stern to port. That rotated our bow to starboard and brought us in closer to the pier than I intended. If it hadn't been for the tug we had made up on the port bow, we could have been in real trouble. As it was, we just didn't look very sharp."

"If we learned something, it was worthwhile," said the skipper. "I completely agree with your analysis—and you did a good job of using the tug to sort things out. The lesson we all need to take away from this is that a whole variety of forces act on the ship, and we need to anticipate, understand, and make use of them. Any questions?"

The ship is suspended in a fluid medium. At very low speeds it is almost frictionless, but as speed increases, the force required to move the ship goes up sharply. It moves in response to the vector sum of all of the forces exerted upon it. Some of these forces are under our control, some are not. A prerequisite to becoming a competent shiphandler is to understand all of these forces, how they affect the ship, and how they interact. For our purposes it is useful to group these forces into those directly controllable from the bridge, those which may be controlled by voice communication to a remote location, and those which are not under our control.

Directly controllable forces include those exerted by the ship's engine or engines, normally through propellers, rudders, and auxiliary power units or thrusters. With knowledge of the appropriate ship's characteristics, the timing and magnitude of these forces is subject to relatively precise control. Forces directly controllable from the bridge only through remote communications include lines, anchors, and tugs. These forces, while controllable, are not subject to the same precision as the engines and rudders. Forces not under our control include wind, current, and channel configuration. These forces need to be understood in order to compensate for them or, sometimes, make use of them (see fig. 2–1). In controlling those elements that are under our control, it is essential that standard commands be used to avoid the possibility of misunderstanding. Chapter 3 is devoted to these standard commands.

Directly Controllable Forces

Engines

The ship's engines, working through the propellers, generate force primarily along the axis of the ship. The water offers a resistance to the ship's motion that is proportional to the square of our speed. At rest very little power is needed to move the ship. If we ring up an ahead bell on our engines, the ship will accelerate to a speed at which the thrust generated through the propellers is balanced by the resistance of the water to the passage of the ship.

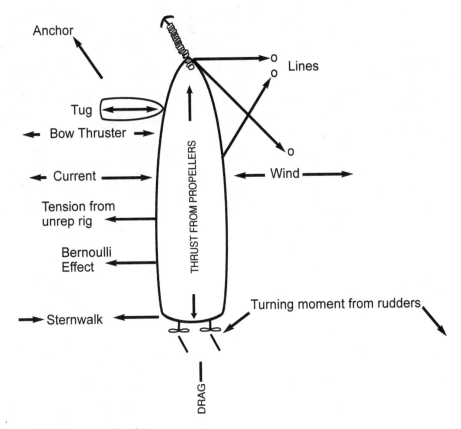

Figure 2–1. Forces on the ship.

Because the resistance of the water to the ship's motion increases so steeply with increases in speed, a ship using half of its power can make about 80 percent of its maximum speed. A destroyer capable of thirty-two knots can typically make twenty knots using only a quarter of its power.

Engines in naval vessels can be gas turbine, steam turbine (either conventional oil fired or nuclear), diesel, or electric. In the case of electric drive, the required electricity can be generated by any of the other power plants. Steam turbines and electric propulsion systems usually drive through fixed-pitch propellers. Gas turbine and diesel plants usually use controllable-pitch propellers. Naval vessels can have one, two, or four shafts. There is no technical obstacle to having three shafts or more than four, but the only known naval examples existing at present are the Coast Guard *Polar Star*– and *Mackinaw*-class icebreakers, which have three shafts.

Figure 2–2. Conventional screw and rudder on USS *Belleau Wood* (LHA 3). *U.S. Navy photo*

Propellers

The propellers are the means of transmitting the power generated by the ship's engines to drive the ship through the water. The pitch of a propeller is the distance its rotation would drive the ship through the water if there were no slippage. Most single-screw propellers rotate in a clockwise direction, as viewed from the stern. Ships with two or four shafts generally have out-turning screws, that is, when going ahead the starboard screw rotates clockwise and the port screw rotates counterclockwise as seen from astern. The *Spruance* and *Ticonderoga* classes, which have their shafts rotating inboard when going ahead, are believed to be the only notable current exceptions to this rule.

Besides the thrust ahead or astern generated by the screw, a side thrust is also generated. The amount of side thrust generated varies greatly from one ship class to another, depending upon the design of the propellers and configuration of the ship's underwater body. It is most notable in single-screw ships, since ships with more than one screw invariably have their propellers rotating in opposite directions. The direction of the side thrust depends upon the direction of the screw's rotation. This may be visualized by thinking of the screw as a wheel resting on the bottom (see fig. 2–3). Sternwalk is more notable in gas

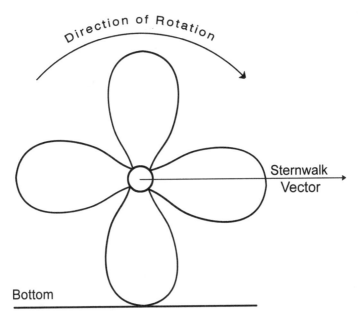

Figure 2–3. The direction of the sternwalk vector created by propeller rotation is as though the propeller were a wheel resting on the bottom.

turbine ships than in others, because the shafts continue to turn even when a stop bell is ordered. When the engines are ordered to stop the variable-pitch propellers are set for minimum pitch, but the shafts continue to rotate and thus induce sternwalk.

With a gas turbine twin-screw vessel with controllable-pitch propellers, sternwalk is not an issue, since the shafts continue to turn in opposite (and counterbalancing) directions even when one engine is ahead and the other astern. With other propulsion systems, the rotation of propellers and shafts is reversed when the engines are ordered astern. If engines are opposed, both shafts will turn in the same direction, generating a sternwalk vector. Since this sternwalk vector is in the same direction as that generated by the lateral twisting force generated by the opposed engines, it is usually unnoticeable, but does assist the ship to twist.

It is with single-screw vessels that sternwalk demands the shiphandler's attention. Most single-screw vessels have propellers that rotate clockwise as viewed from astern. With a single-screw gas turbine ship such as the *Oliver Hazard Perry*–class frigate, whose shaft and propeller continue to rotate in the same direction whether set for ahead, stop, or back, this generates a notable tendency for the stern to walk to starboard under all circumstances unless balanced by the auxiliary propulsion units. This tendency must be taken into account in planning any evolution. Most other single-screw ships reverse the propeller and shaft when backing. These ships generate a side-walk vector that moves the stern to port when backing. In contrast to the variable-pitch ship, however, no sternwalk vector is generated when the engines are ordered to stop, since the shafts and propellers cease to rotate.

Rudders

The first thing to understand about rudders is that they generate a force on the ship only when water is flowing past them. The flow past the rudders can be generated by the motion of the ship, the discharge from the propellers, or both. Ships have one or two rudders. They are always located aft of the propellers so that when the propellers are turning ahead, their wash is directed at the rudders. This means the rudders will always be more effective when going ahead than when going astern. Control of a backing ship is almost always less precise than when going ahead.[1]

The rudder, located well aft on the hull, works by moving the stern of the ship in the opposite direction from the intended turn. To turn right, the rudder swings to the right. This creates a pressure differential on the rudder, with higher pressure to starboard, reduced pressure to port. The resulting

Figure 2–4. Controllable-pitch propeller, here on USS *McInerney* (FFG8). *J. Bouvia*

force moves the stern to port. This creates a drift angle creating pressure on the port bow of the ship. The pressure on the port bow combined with the rudder force pushing the stern to port causes the ship to turn to starboard. Note that this force generated by the rudder causes the ship to move initially slightly off track in the direction opposite to the turn (see fig. 2–6).

Figure 2–5. Rudders generate a force only when water is flowing past them. Note the way in which a wash from the propeller will be directed at the rudder. *U.S. Naval Institute photo archives*

The turning force generated by the rudder depends upon the rudder's size, angle, and the velocity of the water moving past it. The velocity of the water in turn depends upon both the motion of the ship through the water and the velocity of the propeller wash generated by the ship's screws. Ships with rudders located directly behind the propellers are more responsive than those with offset rudders (as, for example, a twin-screw ship with a single rudder). For a stopped ship, an ahead bell against the rudder can generate substantial turning force before the ship gathers much way. Thus the tightest turns can be made from a dead stop.

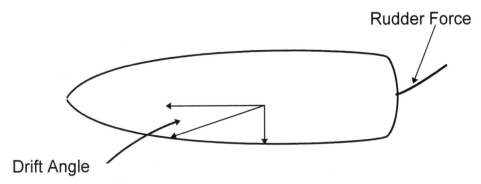

Figure 2–6a. Right rudder for a starboard turn creates a force to port. The result of the ahead force and the rudder force to port creates a drift angle, initially moving the ship to the left of the track.

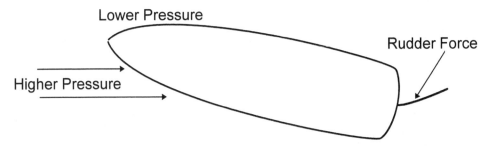

Figure 2–6b. As the stern moves to port, hydrodynamic pressure on the port bow, combined with reduced pressure on the starboard bow, turns the ship to starboard.

A rudder generates both turning force and drag. Both increase as the angle of the rudder increases, but not in a linear way. The amount of turning force is incrementally less as the angle increases, while the amount of drag is incrementally more (see fig. 2–7). Because of the possibility of damage if a rudder is slammed into its stops, it is best to limit hard rudder to emergencies, or to first order full rudder, then move the rudder gently the last few degrees to hard rudder.

The rudder is less effective when going astern than when going ahead. The propellers turning astern do not generate an effective flow across the rudder. Thus the rudder does not become effective when going astern until significant sternway is gathered. This varies substantially with ship type, but most ships do not steer reliably astern much below five knots of sternway. It is important for the shiphandler to understand that the disturbed flow a backing bell creates for the rudders can negatively affect steering, even while

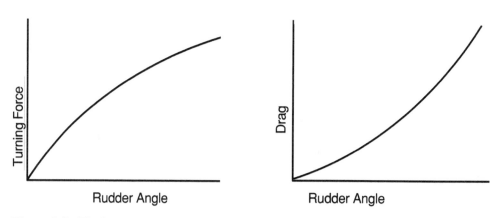

Figure 2–7. The increase in turning force becomes relatively less with each degree of increase in rudder angle, and the increment of drag generated by the rudder becomes relatively greater.

the ship still has substantial headway. In some circumstances, you have to choose between being able to stop or being able to steer, but not both.

It is important to know how your ship responds to various angles of rudder at various speeds. Two important aspects of this are called advance and transfer. Advance is the distance that the ship will move forward in a 90-degree turn in response to a specified rudder angle. Transfer is the lateral distance the ship will move during the same maneuver. Tactical diameter is the lateral distance the ship will move in a 180-degree turn with a specified rudder angle. Standard tactical diameter is specified so that ships in formation will turn at the same rate during tactical maneuvers (see fig. 2–8).

The Moving Pivot Point

In turning, the ship rotates about a pivot point. Forward of that point the ship when turning moves in one direction, and aft of the pivot point moves in the opposite direction. Any ship's pivot point is not fixed; it varies with circumstances. It is not unusual to be told something like "On this ship the pivot point is slightly aft of the bridge." Sometimes it is, sometimes it is not. What is really being said is something like "When this ship is going ahead with substantial way on, and has no tugs, lines, or anchors exerting forces, it will respond to the rudder by rotating about a point slightly aft of the bridge." In fact, the pivot point can move over almost the entire length of the ship as circumstances vary.

A ship's center of lateral resistance is determined by the shape and area of its underwater body, and is the point at which there is as much underwater area forward as there is aft. If a lateral force is applied to the ship at this point,

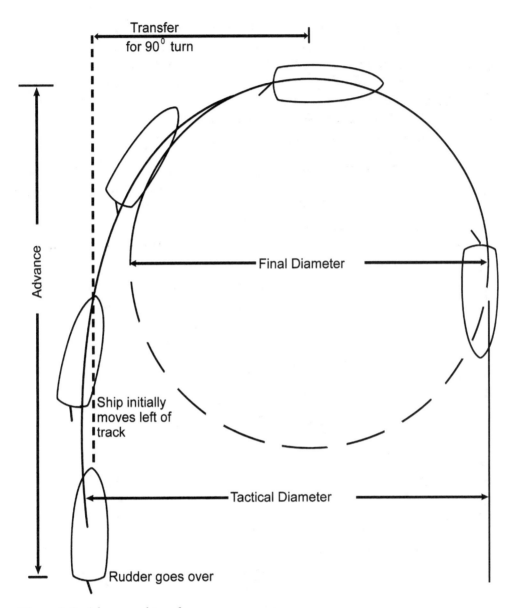

Figure 2–8. Advance and transfer.

as by a tugboat, the ship will move sideways without rotating. If a lateral force is applied forward or aft of this point, the ship will rotate in addition to moving sideward. The pivot point of this rotation will be offset from the center of lateral resistance in a direction opposite to the offset of the applied force. If the tug pushes near the stern, the pivot point will be toward the bow. The further the applied lateral force is from the center of lateral resistance, the further the pivot point will move in the opposite direction, and the more the applied force will tend to rotate the ship rather than move it laterally. Thus if you want a tug to move a ship laterally without rotation, it should be made up amidships near the center of lateral resistance. If you want the tug to rotate the ship, it should be placed near the bow or stern. See figure 2–9, which illustrates how the pivot point moves in various circumstances.

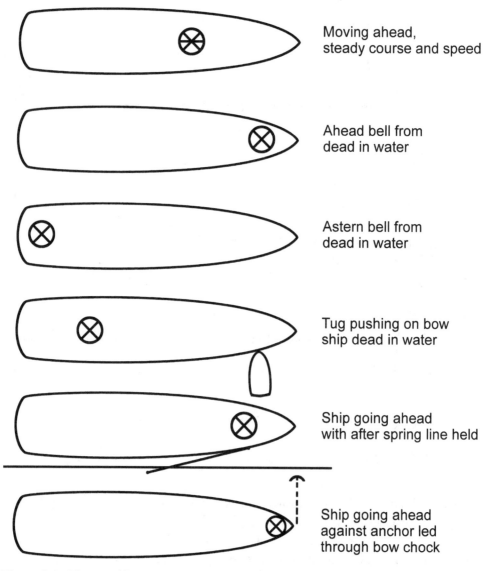

Figure 2–9. The movable pivot point.

The use of mooring lines or the anchor also can move the pivot point. If the ship moves ahead against the anchor, the pivot point moves to the bow. Under these circumstances the ship's engine(s) and rudder may be used to obtain precise positioning of the stern, while the anchor holds the bow in place. Similarly, a spring line (a mooring line that makes an acute angle to the ship and to the pier) can be used to move the pivot point. The most frequent use of this is in moving the stern away from the pier to get under way. To do this the ship's engines are worked ahead against a mooring line run from the bow to a place well aft on the pier. (The correct nomenclature for this line is "forward after spring line," usually line number two.) This moves the pivot point well forward, so that the propeller discharge against the rudder full over toward the pier moves the stern away from the pier while the bow remains in place. The line in this case serves the dual purpose of moving the pivot point well forward and restricting the ship's forward motion. Line-handling terminology is covered in chapter 3. Diagrams of standard mooring line configurations for destroyers and aircraft carriers are provided in figure 2–10.

When the ship is moving though the water it encounters resistance from the water. As is the case with lines or anchor, this resistance moves the pivot point away from the center of lateral resistance toward the retarding force. Thus a ship moving forward will have a pivot point well toward the bow. When going astern the pivot point moves toward the stern. The moving pivot point can be used to advantage when maneuvering in tight quarters. When the ship is dead in the water, and an ahead bell is used against the rudder, the pivot point of the ship moves well forward, amplifying the effectiveness of the rudder in moving the stern.

Auxiliary Power Units and Thrusters

Count yourself fortunate if your ship is equipped with auxiliary power units (APUs) or thrusters. Properly used, they permit even a fairly large vessel to carry out all needed maneuvers without the use of tugs, so long as the wind and current are within their power limitations. The basic difference between the two is that APUs are trainable while thrusters are not. As the name indicates, the principal purpose of the APU, such as found on *Oliver Hazard Perry*–class frigates, is to provide a "get home" capability in the event that the ship's single shaft or propeller is disabled. However, it also is a very useful aid to shiphandling. By opposing the APUs to the ship's engine, it is

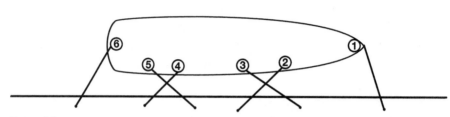

1. Bow Line
2. After Bow Spring Line
3. Forward Bow Spring Line

4. After Quarter Spring Line
5. Forward Quarter Spring Line
6. Stern Line

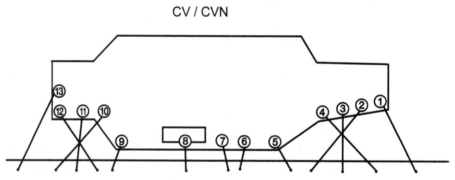

1. Bow Line
2. After Bow Spring
3. Bow Breast
4. Forward Bow Spring
5. Forward Waist Spring
6. After Waist Spring
7. Forward Waist Spring

8. Waist Breast
9. After Waist Spring
10. After Quarter Spring
11. Quarter Breast
12. Forward Quarter Spring
13. Stern Line

Figure 2–10. Standard destroyer and carrier mooring lines.

possible to walk the ship sideways into a slip. The bow thruster, although it cannot be opposed to the ship's screws, does provide a means of moving the bow to port or starboard under precise control. Both the APUs and the bow thrusters lose effectiveness when the ship is proceeding at any significant speed through the water. The use of both of these aids will be discussed in greater detail later on.

Indirectly Controllable Forces

Mooring Lines

A ship's lines are, of course, used to hold the ship securely while moored. They also are an important element of shiphandling. Lines are numbered in order from the bow with regard to their position on the ship. Figure 2-10 shows typical lines as used by a destroyer or cruiser as well as those used by the largest naval vessel, the aircraft carrier. The smaller the vessel, the more use is likely to be made of the lines for shiphandling purposes rather than just for mooring. For larger vessels, the strain that would be placed on the lines could become excessive, so that for these ships the mooring lines are usually used just to hold the vessel in place. The utility of lines is greatly enhanced by line-handling winches fore and aft.

Anchors

The ship's anchors can be used for a great deal more than just tethering the ship to an assigned anchorage. They can serve, within some very important limits, to get us out of tight spots in an emergency, such as loss of propulsion power or rudder control, or where wind and/or current are preventing a desired maneuver. Anchors can also serve as very handy auxiliary means of shiphandling, particularly for single-screw ships. The techniques for doing this are set forth in chapter 6.

Tugs

The larger our ship, and the greater the wind and current, the more important tugs become for handling the ship in restricted waters. Tugs often come as a package with the services of a pilot. The use of tugs and relations with the pilot are discussed in chapter 8.

Uncontrolled Forces

Wind

Although wind is perhaps not now as overriding a factor for the seaman as it was during the days of sail, it remains a major factor in the control of ships. It is perhaps next only to the ship's engines and rudders as a force affecting our ship. Air operations, underway replenishments, under ways and landings, and our ability to twist or turn the ship in restricted waters are all greatly affected by the wind.

The wind is outside our control, and it can change quickly. Weather predictions and the ship's anemometers can give us a preliminary indication of what to expect, but nothing takes the place of immediate observation. The shiphandler needs to learn to read flags and the surface of the water to determine wind direction and velocity. Time spent in a sailboat provides excellent practice in reading the wind and understanding its vagaries. Structures and other ships can cause eddies and can block the wind either partially or entirely. You have probably noticed the flags above a football stadium flying in different directions, with the wind on the field being different yet. The same thing can affect our ship.

Ships vary greatly in their response to the wind. The more sail area the ship presents to the wind, the more it will be affected. A deep-draft ship with relatively low sail area, such as a submarine, will be moved much less by the wind than will a shallow-draft ship with a large sail area, such as an amphibious ship. Most warships have more sail area forward than aft, so that a beam wind pushes more on the bow. Since this pushes the bow down wind more rapidly than it does the stern, a rotating force is developed tending to cause the ship to back into the wind. This tendency makes it more difficult to twist the bow of the ship through the wind.

The wind is often the determining factor in deciding whether tugs are needed when going alongside or getting under way from a pier. It is not possible to state a given wind velocity that requires the use of tugs. If the wind is blowing straight down the pier, we can accept higher winds than if the wind is from the beam. A wind from directly ahead or astern is relatively easy to balance with our engines. A beam wind is more likely to present a problem.

Current

Current is water in motion, and it affects our ship in much the same way as does the wind. Currents can be generated by tides, rivers, or prevailing winds. To add a degree of difficulty for the shiphandler, it is not at all unusual for the wind and the current to come from different directions. The relative importance of wind and current varies from ship type to type. Often a ship will adopt a rule of thumb, such as thirty knots of wind equals one knot of current. As with wind, a current is more difficult to compensate for when it is from the beam than from ahead or astern. A current from ahead is also easier to deal with than one from astern, since the ahead bell used to offset the current improves the effectiveness of our rudder.

As with the wind, the shiphandler needs to know in advance what current to expect and to augment this knowledge with direct observation. You should never take the conn to operate in restricted water without knowing the state of the tide. The *Tidal Current Tables*, which the navigator has, give the velocity of flow and the times of slack water (no flow), maximum ebb (current flowing outward), and maximum flood (current flowing inward). In addition to this you need to check current visually. Currents can be observed on the surface of the water by the visible wakes and eddies from navigation marks, pier pilings, crab pots, and so on (see fig. 2–11). Anchored ships will lie to the current, or the wind, or somewhere in between. Currents tend to cause can and nun buoys to lean down current, although this observation must be made with caution: buoys can also sometimes lean because they have taken on water, or for idiosyncratic reasons of their own.

In the case of both wind and current, there are additional pitfalls of which you need to be aware. Either or both can affect the turning point when con-

Figure 2–11. With a little experience the direction and speed of the current can be accurately estimated by observing buoys and pilings. *U.S. Naval Institute photo archives*

forming to a channel, or require the use of more or less rudder than would be appropriate in their absence. If the wind or current is across the channel, the ship will have to steer an offsetting course (called "crabbing") to remain in the channel. The slower the speed of the ship, the more offset will be required. If there are screening objects such as a building that blocks the wind or a solid face pier that blocks the current, we may find that part of our ship is still exposed while the remainder is protected. This can trip up the unwary shiphandler by imparting an unexpected rotation.

Bernoulli Principle

In 1738, Swiss physicist Daniel Bernoulli published his most famous work, *Hydrodynamica*. In it, he formulated the principle that within a fluid under conditions of steady flow the sum of the energy of velocity, the energy of pressure, and the potential energy of elevation remains constant. This was one of the earliest formulations of the principle of the conservation of energy. One of the most important implications of Bernoulli's principle is that an increase in the velocity of a fluid, whether liquid or gaseous, causes a decrease in pressure. This is the basis on which rest the lift of an airplane wing, the measurement of flow in a pipeline, ground effects in a race car, the ability of a sailboat to sail to windward, and the force of attraction between two ships steaming in parallel. The Venturi effect is a derivative of the Bernoulli principle.

Venturi Effect

Fluid speed increases when the fluid is forced through a narrow or restricted area. The increased speed of fluid flow results in a decrease in pressure. There are several circumstances in which the Venturi effect creates a force on our ship. During an underway replenishment, two or more ships steam on parallel courses in close proximity. As the water flows between the ships, it must pass through a narrower passage, speeding the flow as it does so. The consequence is a reduction in the pressure on the inboard sides of both ships, tending to suck the ships together. The greater the speed of the ships, and the closer they are together, the stronger is the force trying to bring them together. How to take this force into account is covered in more detail in chapter 9 on underway replenishment.

Channel Configuration

Narrow channels or shallow water can contribute forces of their own. Shal-·low water effect can greatly reduce the responsiveness of the ship to its rudder. Shallow water can also create a squatting effect reducing the ship's clearance over the bottom and increasing the power required for a given speed. This is a special example of the Bernoulli principle, in which the water flowing between the sea bottom and the ship's keel is accelerated with a resultant decrease in pressure, causing the ship to sit lower in the water. A narrow channel with steep side gradients can also create both suction and cushioning effects that the shiphandler needs to understand. These are discussed in chapter 7.

The controllable forces listed here are the ship's engines, propellers, rudders, auxiliary power units or thrusters, mooring lines, anchors, and tugs. The forces not under our control are wind, current, Venturi effects, and channel configuration. Having identified these forces, subsequent chapters develop how the shiphandler uses them or compensates for them in order to maneuver the ship to the intended place.

3

STANDARD COMMANDS

————◄◦►————

I n the wardroom of USS *San Antonio* (LPD 17), Ens. Dwight Cunningham was completing his assigned teaching session on standard commands to the rest of the ship's officers. The senior watch officer, Lt. Rosemary Stewart, was playing devil's advocate.

"Dwight, you've told us what standard commands are, but you haven't said why they matter," said Lieutenant Stewart. "Isn't this just another case of military insistence on formality, whether it makes any practical sense or not? Who cares if I say, 'Give me 15 degrees of rudder to starboard' instead of 'Right standard rudder,' so long as the helmsman understands me?"

"It makes a lot of difference," replied Ensign Cunningham. "We use standard commands to minimize the possibility of misunderstanding. Ambiguity and misunderstanding of commands to the rudder, the ship's engines, to the anchors, to tugs, or to the handling of lines can lead to accidents and injuries. I've been impressed by the precision standard commands bring to orders. The more stressful the circumstances, the more important it is to use standard commands. Standard phrasing is a lot easier for people to recognize and understand than orders in an unfamiliar form. I've seen enough garbled phone communications even without complicating things with unfamiliar phrasing."

"Okay, that's what I wanted to emphasize," said Lieutenant Stewart. "But there is another good reason. Everyone in the Navy gets transferred with regularity. We shouldn't have to learn a new set of commands every time we change ships. Fortunately, standard commands apply to commissioned naval vessels, of all types, whether large or small. It can be tempting to develop shortcuts and individual variations, but this is a very bad idea. You've made your point well, Dwight. Thanks."

In classes on leadership, it is usual to identify a span of leadership styles ranging from participative-democratic to autocratic. The instructor generally concludes that there are no circumstances in which autocratic leadership is to be preferred. True as this may be in general, it is not so where conning a ship is concerned. Orders to the helm, lee helm, and line-handling stations are not matters to be discussed. Conning a ship requires autocratic leadership: crisp, sharp, clear, precise, and unambiguous orders. That is why we use standard commands.

In the interest of standardization, the commands set forth in this chapter are, unless otherwise indicated, identical to those published in *Watch Officer's Guide*.[1] Orders must not only be standard but also must be given in a tone and

Figure 3–1. Standard commands minimize the possibility of misunderstanding. *U.S. Navy photo*

loudness that stamps them immediately as a command. Few things are worse than a timid or tentative order. The recipients of the order must be able to understand clearly and without possibility of ambiguity that an order is directed to them. To aid in this, and to avoid the possibility of misunderstanding, orders follow a standard and logical pattern.

Orders to the helmsman start with the direction the rudder is to go: "Right" or "Left." The second word indicates how far the rudder is to be turned. This can be in degrees or a predefined angle, such as "Standard" or "Full." The third word in the command is "Rudder." This sequence permits quick and accurate response, as the helmsman can start the rudder in the proper direction immediately, then stop the rudder swing on the ordered angle. The exception to this is that when maximum rudder is needed, we preface the order with "Hard" to indicate urgency. The order in this case is "Hard right (left) rudder."

Orders to the engine order telegraph (or "lee helm") are constructed in a similarly logical way. The first term, "All engines," or "Port (starboard) engine," tells the operator which handle is to be moved. The next word, "Ahead," "Stop," or "Back," tells in which direction they are to be moved. The last part of the command, "One-third," "Full," and so on, tells the amount of the speed change. Standard commands to the engines include

1. "All engines ahead one-third (two-thirds, standard, full, flank)," "All engines stop," or "All engines back one-third (two-thirds, full)";

2. "Starboard (port) engine, ahead one-third (two-thirds, standard, full)," "Starboard (port) engine stop," or "Starboard (port) engine, back one-third (two-thirds, full)."

Each ship will normally have standard acceleration tables to make the rate of acceleration predictable and uniform between ships of a class. In an emergency, when we want maximum acceleration or deceleration, the command is, "All engines ahead (back) emergency." In response to this command, the operator rings up "Ahead flank" (or "Back full") three or more times in rapid succession, ending with the handles at the flank or back full position.

In addition to the handles on the engine order telegraph that indicate desired speeds in five-knot increments, there is also a revolution indicator. When quick response in five-knot increments is desired, this is indicated by ringing up on the rpm indicator a speed well beyond the engines' range, such as "999." The order for this is "Indicate 999 revolutions for maneuvering bells." When a specific speed is desired, the exact number of revolutions to be made is indicated on the revolution indicator, and the corresponding speed (one-third, two-thirds, etc.) on the handles of the engine order telegraph. An example would be "All engines ahead standard, indicate 240 revolutions for sixteen knots." The term "revolutions" must always be included in these orders to prevent possible confusion with orders concerning course or bearings. When changing speed in small increments, resist the temptation to use terms such as "Add three turns" or "Take off two." Instead, say the exact number of revolutions to be rung up, as in "Indicate one one seven revolutions." If the speed change takes you across a division such as between one-third and two-thirds, the order will be "All engines ahead two-thirds, indicate one one seven revolutions."

If the conning officer is in a position from which he is unable to see or remember the revolutions required, he can substitute the command "Indicate turns for ——— knots," requiring a report back of the turns actually rung up, as well as a repetition of the command.

The report back after every order to the rudder or engines is an essential part of the system of standard commands to ensure that the command is properly understood. It also serves as a check on the officer giving the command to make sure that he has not given an incorrect command through a slip of the tongue. The final step in the system is a report from the helmsman or lee helmsman that the order has been carried out. This is acknowledged by the conning officer with a "Very well."

Steering Commands to the Helm

The following examples of steering commands to the helm are taken directly from the *Watch Officer's Guide* (14th ed., pp. 142–49):

> When a specific amount of rudder is desired:
> Command: "Right full rudder (or right standard rudder)."
> Reply: "Right full (standard) rudder, aye, Sir (or Ma'am)."
> Report: "My rudder is right full (standard), Sir (or Ma'am)."
> When the rudder order is given in degrees:
> Command: "Left ten degrees rudder."
> Reply: "Left ten degrees rudder, aye, Sir (or Ma'am)."
> Report: "My rudder is left ten degrees, Sir (or Ma'am)."
> (*Note:* When a rudder order has been given, but no course to come to, the helmsman should report passing each ten degrees: "Passing 270, Sir (or Ma'am)." Besides the information provided, this serves as

a reminder to the conning officer that no course order has yet been given. Each report should be acknowledged with a "Very well." The conning officer may direct "Belay your headings" if he does not desire the reports, but this should be done with caution since it removes a useful reminder.

When the helmsman is to steady on a specific course:
    Command: "Steady on course ———."
    Reply: "Steady on course ———, aye, Sir (or Ma'am)."
    Report: "Steady on course ———, Sir (or Ma'am). Checking ——— magnetic."

When maximum possible rudder is required:
    Command: "Hard right rudder."
    Reply: "Hard right rudder, aye, Sir (or Ma'am)."
    Report: "Rudder is hard right, Sir (or Ma'am)."
    (*Note:* The danger in using hard rudder lies in the possibility of jamming the rudder into the stops. For this reason it is rarely used except in emergencies. If hard rudder is chosen when there is no emergency, the conning officer may reduce the possibility of jamming the rudder by first ordering full rudder and then increasing the rudder to hard, allowing the helmsman more control of the rudder's movement.)

When the amount of rudder is to be increased:
    Command: "Increase your rudder to ——— (right full, right ten degrees, etc.)."
    Reply: "Increase my rudder to ———, Sir (or Ma'am)."
    Report: "My rudder is ——— (right full, right ten degrees, etc.)."

When the amount of rudder is to be decreased:
    Command: "Ease your rudder to ——— (standard, left ten degrees, etc.)."
    Reply: "Ease my rudder to ———, Sir (or Ma'am)."
    Report: "My rudder is ——— (standard, left ten degrees, etc.)."

When the rudder is increased or decreased while the ship is turning to an ordered course:

Command: "Right standard rudder, steady on course 270."

Reply: "Right standard rudder, steady on course 270, aye, Sir (or Ma'am)."

Command: "Increase your rudder to right full, steady on course 270."

(*Note:* When the rudder is increased or decreased, the conning officer must restate the desired course. Otherwise the helmsman is to leave his rudder at the position last ordered and report passing each ten degrees until a new course is given.)

Reply: "Increase my rudder to right full, steady on course 270, aye, Sir (or Ma'am)."

Report: "My rudder is right full, coming to course 270, Sir (or Ma'am)."

When course change is less than ten degrees:

Command: "Come right steer course ———."

Reply: "Come right, steer course ———, aye, Sir (or Ma'am)."

Report: "Steady course ———, checking course ——— magnetic, Sir (or Ma'am)."

A command once standard, now obsolete, is "Come right (left) handsomely to ———." Its meaning is to do so slowly and carefully, but it has been dropped because easily misunderstood as meaning the opposite.

When the rudder angle is to be reduced to zero:

Command: "Rudder amidships."

Reply: "Rudder amidships, aye, Sir (or Ma'am)."

Report: "My rudder is amidships, Sir (or Ma'am)."

When the course to be steered is that which the ship is on at the instant the command is given:

Command: "Steady as you go."

Reply: "Steady as you go, course ———, aye, Sir (or Ma'am)."

Report: "Steady on course ———, Sir (or Ma'am), checking ——— magnetic."

(*Note:* This command should normally not be used if the ship's head is swinging rapidly. Injudicious use could cause momentary loss of control over the ship's swing if the helmsman is required to use a large rudder angle to carry out the order. To prevent this, the order should be preceded by "Rudder amidships." This, of course, requires anticipation on the conning officer's part to ensure a correct heading.)

When the swing of the ship is to be stopped without steadying on any specific course:
   Command: "Meet her."
   Reply: "Meet her, aye, Sir (or Ma'am)."
   (*Note:* Immediately after the reply is given, the conning officer must order a course to be steered.)
   Command: "Steady on course ———."
   Reply: "Steady on course ———, aye, Sir (or Ma'am)."
   Report: "Steady on course ———, Sir (or Ma'am). Checking ——— magnetic."

When equal and *opposite* rudder is desired relative to that previously ordered:
   Command: "Shift your rudder."
   Reply: "Shift my rudder, aye, Sir (or Ma'am)."
   Report: "Rudder is ——— (a rudder angle equal but opposite to that previously ordered)."

When the heading of the ship is to be determined at a given moment:
   Command: "Mark your head."
   Report: "Head is (exact heading at that moment), Sir (or Ma'am)."
   (*Note:* If the helmsman appears to be steering properly but the ship is not on its correct heading, the conning officer should use this command to compare the helmsman's compass with other repeaters on the bridge.)

When the helmsman has been given a course to steer but appears to be steering badly or is continually allowing the ship to drift from the ordered course:
   Command: "Mind your helm."
   Reply: "Mind my helm, aye, Sir (or Ma'am)."
   (*Note:* No report necessary.)

When the ship is in a situation where minor deviation from an ordered course may be permitted to one side, but none may be permitted to the other side (for example, when alongside another ship for refueling):

Command: "Steer nothing to the left (right) of course ———."

Reply: "Steer nothing to the left (right) of course ———, aye, Sir (or Ma'am)."

(*Note:* No report necessary.)

It is generally poor practice to order a course to steer when the ship is maneuvering, since this relinquishes control from the conning officer to the helmsman. By giving rudder orders the conning officer retains control. It makes no sense to give a course to steer when the ship is maneuvering in a slip, when engines are backing, when the ship has no way on, or when twisting. Best practice is for the conning officer to bring the ship to the desired course through rudder orders, and only then relinquish control to the helmsman by giving him a course to steer. This is also excellent practice in learning precise control of your ship.

Whenever ordering a course change, the conning officer should perform the following activities:

1. Check the side to which he or she intends to turn to make sure that it is safe to turn in that direction.

2. The ship's speed determines how quickly its head will swing. At very low speeds, a large angle of rudder may be required to bring about a course change. At very high speeds, a large rudder angle may cause it to swing so rapidly that it cannot be safely controlled. All conning officers should be familiar with the tables of turning speeds and turning diameters for their ship. This information is contained in the ship's tactical data book. Generally, the sum of rudder order plus speed in knots should not exceed thirty or else there will be a probability of fairly heavy rolls. On the other hand, at very slow speeds it is appropriate to use large amounts of rudder: full or even hard rudder at speeds below five knots.

3. After giving a rudder order, the conning officer should monitor its execution by checking the rudder angle indicator to ensure there was no misinterpretation of the command.

Engine-Order Commands to the Lee Helm

Engine orders are always given in the following order:

1. Engine: Which engine is to be used. If both engines are to be used, the command is "All engines." On single-screw ships the command is always "Engine."

2. Direction: Ahead, back, or stop.

3. Amount: Ahead one-third, two-thirds, full, flank. Back one-third, two-thirds, full.

4. Shaft revolutions desired: Number of revolutions in three digits for the desired speed in knots. (Shaft revolutions are not used for backing orders.)

The following are examples of commands to the lee helm:

When a twin-screw ship is to go ahead on both engines to come to a speed of six knots:

Command: "All engines ahead one-third. Indicate zero eight eight revolutions for six knots."

Reply: "All engines ahead one-third. Indicate zero eight eight revolutions for six knots, aye, Sir (or Ma'am)."

Report: "Engine room answers all ahead one-third. Indicating zero eight eight revolutions for six knots, Sir (or Ma'am)."

When different orders are given to port and starboard engines, revolutions should not be specified:

Command: "Port engine ahead one-third, starboard engine back one-third."

Reply: "Port engine ahead one-third, starboard engine back one-third, aye, Sir (or Ma'am)."

Report: "Engine room answers port ahead one-third, starboard back one-third, Sir (or Ma'am)."

When the order is to only one engine, the report must include the status of both engines:

Command: "Starboard engine ahead one-third, port engine back one-third."

Reply: "Starboard engine ahead one-third, port engine back one-third, aye, Sir (or Ma'am)."

Report: "Engine room answers starboard engine ahead one-third, port engine back one-third, Sir (or Ma'am)."

Command: "Starboard engine stop."

Reply: "Starboard engine stop, aye, Sir (or Ma'am)."

Report: "Engine room answers starboard engine stop. Port engine back one-third, Sir (or Ma'am)."

When there are to be small changes in speed, for example when the ship is alongside another for refueling or to keep station on the formation guide:

Command: "Indicate one zero zero revolutions."

Reply: "Indicate one zero zero revolutions, aye, Sir (or Ma'am)."

Report: "Engine room answers one zero zero revolutions for three revolutions over eleven knots, Sir (or Ma'am)."

On many gas-turbine ships with controllable reversible-pitch propellers, at speeds below twelve knots (the exact break point varies between classes), the ship's speed is controlled by varying the pitch of the propeller blade, measured as a percentage. This requires additional orders at lower speeds, as in:

Command: "All engines ahead one-third. Indicate ——— revolutions, ——— percent pitch for ——— knots."

Reply: "All engines ahead one-third, indicate ——— revolutions, ——— percent pitch for ——— knots, aye, Sir (or Ma'am)."

Report: "All engines answer one third. Indicating ——— revolutions, ——— percent pitch for ——— knots, Sir (or Ma'am)."

(*Note:* This unfortunately will vary from ship to ship, even within a class. Check the captain's standing orders. It is time for the type commanders to convene a panel to recommend and promulgate new appropriate standard commands for ships with adjustable pitch propellers.)

When maneuvering in restricted waters, getting under way, docking, or mooring, ships usually use what are known as maneuvering bells. Under these circumstances, only engine, direction, and amount are given. Revolutions are not specified. Depending on the type of ship, each engine amount is equivalent to a standard number of knots, the most often used example being one-third equals five knots, two-thirds equals ten knots, and so on. When maneuvering bells (or "maneuvering combination") are desired, the conning officer must order as follows:

Command: "Indicate maneuvering combination (bells)." (By convention, this is usually an engine order for nine nine nine revolutions.)

Reply: "Indicate (nine nine nine) revolutions for maneuvering combination (bells), Sir (or Ma'am)."

Report: "Engine room answers (nine nine nine) revolutions for maneuvering combination (bells), Sir (or Ma'am)."

Line Handling

Standard commands to line handlers are as important as are rudder and engine orders. In a sense they may be even more important, since orders to line handlers are typically transmitted through phone talkers or handsets. This means that the conning officer is often not in a position to observe whether the order has been correctly understood and carried out. Thus it is of particular importance that line-handling commands be standard to minimize confusion and misunderstanding. Many an excellent landing has been ruined by line-handling problems.

Orders to line handlers begin with the action to be carried out, followed by an identification of the line or lines by number, as necessary. Lines are

Figure 3–2. Standard commands apply to line handling as well as to all other aspects of ship control. *U.S. Navy photo*

numbered from bow to stern in the order in which they are attached to the ship (see fig. 2–6). Lines which run perpendicular to the ship are breast lines. Lines which run at an angle from the ship to the pier are spring lines: those which tend aft are after springs; lines which tend forward are forward springs. If an amidships breast line is used, it is not assigned a number. Typical lines for a destroyer are

Line 1: bow line (usually a breast line, or led somewhat forward);
Line 2: after bow spring line;
Line 3: forward bow spring line;
Line 4: after quarter spring line;
Line 5: forward quarter spring line;
Line 6: stern line (usually a breast line, or led somewhat aft).

Standard commands to the lines, following those set forth in *Watch Officer's Guide*, are as follows:

Command	Meaning
"Stand by your lines"	Man the lines, ready to put them over, cast them off, or take them in.
"Take in all lines"	Slack off lines and signal people tending lines on the pier or on another ship to cast off our lines (men of war normally use their own lines. On occasion lines may be taken from the pier or another ship. In this case, the command to return lines will be "Cast off all lines" or by number if appropriate, as in "Cast off line 3)."
"Over all lines"	Pass the lines to the pier or another ship, place the eye of each over the appropriate bollard but take no strain.
"Take a strain on (line 3)"	Put the line under tension.
"Slack (line 3)"	Take tension off the line, and let it hang slack.
"Ease (line 3)"	Let out enough of the line to lessen tension.

"Take (line 3) to the capstan or to power"	Lead the end of the line to the capstan, take the slack out of it, but put no strain on it without further orders.
"Heave around on capstan or to power"	Apply tension on the line with the capstan. Follows the order to take the line to the capstan.
"Avast heaving"	Stop the capstan.
"Hold what you've got on (line 3)"	Hold the line as it is.
"Hold (line 3)"	Do not allow any more line to go out. ("Hold" commands should be used with extreme caution because they require the lines to be held even to parting.)
"Check (line 3)"	Hold heavy tension on the line, but let it slip as necessary to keep it from parting.
"Surge (line 3)"	Hold moderate tension on the line, but let it slip enough to permit the ship to move.
"Double up (all lines)"	Pass additional bights on all mooring lines so that there are three parts of each line to the pier or the ship alongside.
"Single up (all lines)"	Take in all bights and extra lines, leaving only a single part of each of the normal mooring lines.
"Cast off all lines"	Used when secured with *another* ship's lines in a nest. Cast off the ends of the lines and allow the other ship to retrieve its lines.
"Shift"	Used when moving a line along a pier. Followed by specification of the line and where it is to go, as in "Shift no. 3 from the bollard to the cleat."

Tug signals may be found in chapter 9 and appendix B.

The Future

For more than a century, the U.S. Navy has used a verbal process of conveying ship control orders to rudder and engines. With very little change over the years, the oral process provides (1) order, (2) acknowledgment of the order, and (3) completion/accomplishment of the order. A Navy officer with the conn learns, very early in shipboard time, that the way to change direction of ship's movement is to order "Right full rudder." The helmsman then acknowledges receipt of the order by repeating, "Right full rudder, aye, Sir (Ma'am)" and turning the wheel to the proper indication. When the rudder reaches the proper position as displayed by the rudder angle indicator, the helmsman reports completion of the order with "The rudder is right full, Sir (Ma'am)," and the conning officer responds, "Very well."

To vary ship's speed, the conning officer orders, "All engines ahead standard, indicate turns for sixteen knots." The lee helmsman acknowledges receipt by repeating the order and adding, "Aye, Sir (Ma'am)." When the console indicates that the order has been carried out, the lee helmsman reports, "All engines are ahead standard, indicating 83 rpm for sixteen knots, Sir (Ma'am)." The conning officer responds, "Very well."

It is a long process of many precisely spoken words, long proven effective, reliable, and useable under all conditions of environment, including combat. Over the past century every individual who has served in and handled a Navy ship has learned and used this well-established process. As long as verbal/oral orders are used, it is most likely that the process of order, acknowledgment, and completion will be retained, even with the Integrated Bridge System.

One portion of the verbal/oral order process of ship control that may change in new ships is the method of ordering ship's speed or engine/propeller thrust. New ships are being equipped with ship control consoles including thrust indicators using Programmed Console Logic (PCL).

The *Arleigh Burke*–class (DDG-51) destroyer ship control console utilizes PCL numbers from 0.0 to 10.0 for speeds from 0 to more than thirty knots. Logic built into the controls sets the appropriate rpm and propeller pitch for the desired speed. For example, if a conning officer desires sixteen knots he/she could order, "All engines ahead five point two." The console operator would set PCL at 5.2 and the ship would develop sixteen knots with 71 rpm and 100 percent pitch on both engines. To produce the equivalent of "All engines back two-thirds," a conning officer could order, "All engines back one point eight." The console would be set –1.8 and the ship would develop ten knots of reverse speed with 69 rpm and negative 45 percent pitch. There are separate PCL indicators for each engine, so that the port engine can be set at 1.2 and the starboard engine –0.6 for the equivalent of "Port engine ahead one-third. Starboard engine back one-third."

In most common use, the conning officer still speaks the traditional language of ship control—ahead, back, one-third, two-thirds, full, and so on—and the lee helmsman or console operator dials in the appropriate PCL number. Each ship has a speed chart used by the console operator to convert traditional orders to PCL settings, and in the future conning officers may make the conversion and give thrust orders by PCL numbers.

4

GETTING UNDER WAY

—◄◦►—

"**N**ow go to your stations all the special sea and anchor detail" sounded over the 1MC. The crew of USS *McClusky* (FFG-41) hustled to their stations in preparation for going to sea. *McClusky* was scheduled under way at 0845, but some had been preparing for the event even before the word was passed. Ens. Jim Porter had gone to the bridge/pilot house area half an hour before sea detail was passed to go over the procedures that he intended to use and to review conditions.

Two days before at the voyage planning meeting, Jim had been told that he would get the ship under way and take it out the channel. He had listened carefully as the navigator briefed the channel configuration and as details of line handling had been given by the first lieutenant. There had been information on expected weather conditions including visibility, tide, current and wind, and the chief engineer reported the propulsion plant readiness. Jim took notes in his little green "wheel book" to ensure that these details would be available when he needed them.

As soon as that meeting was over Jim went to the ship's bridge and studied NOAA Chart 18773, San Diego Bay, on the navigator's table. With dividers he measured distance between piers, length of Pier 7 and distance from the buoy line to shoal water, and wrote these in his wheel book along with depths of water. He compared data with the *Fleet Guide* and recorded additional information on the pier area and the channel that he would take to get *McClusky* to sea. He noted that a pilot was required for more than one tug and smiled, knowing that

Figure 4–1. Guided missile frigate USS *Klakring* (FFG 42). *U.S. Navy photo*

except in unfamiliar ports Commander Lynch, *McClusky*'s captain, never asked for a pilot. From the chart, he obtained the length and true bearing of each leg of the channel. He would use each bearing as his heading, and with the prescribed speed he calculated the time of each leg. Comparing this with data he had recorded from the navigator's brief, he felt comfortable that he knew the channel. Applying expected wind and current, he made notes for each leg of the harbor channel transit.

It had been some time since he had used a *Tidal Current Table,* so Jim asked a quartermaster for help, and the two of them calculated tidal conditions that would exist when *McClusky* was scheduled to get under way, at 0845 on Wednesday. They confirmed their predictions by going to the NOAA web site and Jim smiled. "I guess we got it right, thanks," he said. "Our calculations agree with the navigator's brief."

"They should," replied the quartermaster. "I did it with him, too."

The next morning at 0845, twenty-four hours before scheduled underway time, Jim was again in the pilot house. Last night he had studied his shiphandling book, focusing on the "getting under way" chapter. He had read it before but with a real opportunity imminent, wanted to refresh his memory. This morning Jim verified his data and was pleased to find no changes. He recorded the temperature, barometric pressure, wind direction, and velocity and observed tidal conditions. Yes, slight flood current; he watched a slight eddy across and around the bow of a nearby ship, just what he had expected.

Jim noted that there were two cruisers moored starboard side to, bow in, in a nest astern of *McClusky* on the north side of Pier 7. Across the slip, two LSDs were moored starboard side to, bow out, one astern of the other, so that they occupied the full south side of Pier 6 (see fig. 4–2). He knew that there was 650 feet of water between the piers, but the beam widths of the two cruisers and one amphib plus their camels and fenders would leave a passage of about 430 feet as he backed out of the slip. Plenty of room. He checked the wind again and the current as he surveyed the shiphandling situation. In twenty-four hours this could all change, but thinking it through now was a helpful exercise. With luck, conditions would be the same or very similar tomorrow.

Jim stood on the starboard bridge wing looking at bow, then stern, moving left thumb to indicate auxiliary power unit, right hand for rudder then engine, talking to himself. He twisted his body for hull movement, looked up at a signal flag to read the wind, then started over. Two signalmen walked by and smiled as Jim looked in his wheel book and went back to hand and body movements as he worked out a number of shiphandling procedures. Tuesday night he reread his shiphandling book.

On the bridge Wednesday morning, half an hour before sea detail, Jim went over his data. Wind looked the same, confirmed by the anemometer. Tide and current appeared consistent with his earlier calculations and yesterday's observations. Astern the two cruisers were still nested, and across the slip were the amphibs. Conditions were the

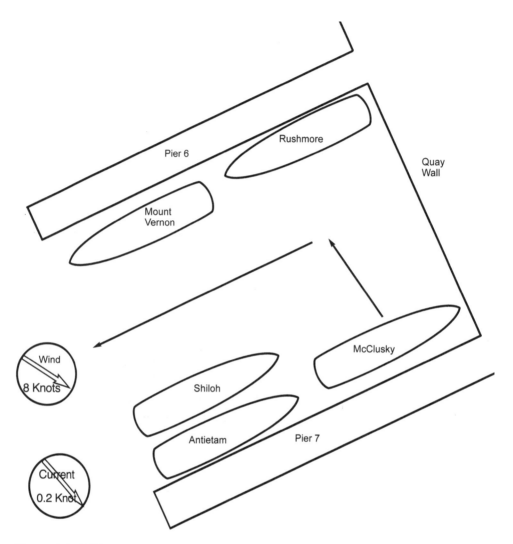

Figure 4–2. USS *McClusky* (FFG 41) under way from Pier 7.

same as twenty-four hours ago. He was studying the chart again as sea
detail was passed and the full bridge/pilot house team started to take
positions. Manned and ready reports were being made and checked off
by the junior officer of the deck (JOOD), and engines were tested.

Twenty minutes after sea detail had been passed, the JOOD went
over the check-off list with Jim. "All stations manned and ready. The
ship is ready to get under way," Jim reported with a salute to the executive officer, who replied, "Very well." Captain Lynch came onto the
bridge and the exec made the same report.

The captain turned to Jim. "Jim, you'll be getting the ship under way this morning," a half question to which Jim responded, "Yes, Sir."

"Will this be your first?"

"Yes, Sir, but I've done it five times in the simulator and with very similar scenarios."

"Good," the captain said, smiling. "Explain to me how you're going to do this."

Jim described the process that he had been going over in his mind the past two days, explaining wind and current and his intended use of mooring lines, engine, rudder, and auxiliary power units. "I'm going to use a five-phase process that starts with Phase 1, 'hovering' the ship alongside the pier with engine ahead and APUs at 180, balancing them so that the ship is stabilized with no headway or sternway. Phase 2 is 'dancing' the ship laterally away from the pier using engine and rudder to move the stern to port and APUs to move the bow to port. When the ship has moved into the center of the slip, it'll be time for Phase 3, backing out into the channel while staying clear, upwind, of the cruisers. Phase 4 is a twist to the channel heading, and Phase 5 is the channel transit to sea."

"How about wind and current?" the captain asked. Jim described wind 8 knots from 290, flood current 0.2 knot toward 130, both setting the ship onto the pier. The amphibs would probably provide some windbreak. Captain Lynch asked how Jim obtained his wind, tide, and current information and predicted conditions for the morning. They discussed use of APUs together or separately, if the APU-engine-rudder combination would be sufficient to overcome effects of wind and current, and the amount of space between cruisers and amphibs in the slip. Jim had all the answers without referring to his wheel book.

"Are you going to use that tug?" Captain Lynch asked, nodding toward the orange and beige tractor tug boat waiting alongside the quay.

"I don't plan on using him. I've done a radio check with him, and he'll just be standing by in case we get in trouble."

"Securite, securite, securite." The bridge-to-bridge radiotelephone channel 16 crackled with a loud clear report announcing *Honda Maru* would be getting under way from National City Marine Terminal in five minutes, outbound to sea. Jim looked at the captain, who turned to the navigator. "How much time for him to get to Pier 7?"

"He's at the Twenty-fourth Street Marine Terminal pier, one mile, with time to get off the pier and up to ten knots, about ten minutes. He should be clear, ahead of us."

The captain turned to Jim and said, "Get under way on time." Jim acknowledged with an "Aye aye, Sir," turned to the phone talker, and ordered, "Single up all lines." Then, "Lower the port and starboard APUs, and train both to 180."

"Securite, securite, securite." USS *McClusky* announced via channel 16 that she would be getting under way from Pier 7 at 0845.

At 0843 Jim ordered, "Central Control Station, stand by to answer bells" and "Fo'c'sle, fantail, stand by your lines." Both orders were acknowledged. "Sound one prolonged blast," and a five-second deep whistle pierced the morning quiet as Jim walked out of the pilot house to the starboard bridge wing.

"Fo'c'sle, bridge, take in lines 1 and 3. Fantail, bridge, take in lines 4, 5 and 6." Both stations acknowledged, and Jim leaned over the combing to see the deck force working lines. Fo'c'sle reported lines 1 and 3 on deck and fantail reported lines 4, 5, and 6 on deck. Jim glanced at the rudder angle indicator on the bridge wing to confirm that the rudder was amidships. "Start Port APU; engine ahead one-third, indicate three knots." When both orders had been acknowledged and carried out, Jim ordered, "Start starboard APU." Lee helm acknowledged then reported, "Port and Starboard APUs trained 180 and on."

"0845," reported the JOOD, and Jim ordered, "Fo'c'sle, bridge, take in line 2." When fo'c'sle reported line 2 on deck, a police-like whistle screamed and the bos'un mate called, "Shift colors," as *McClusky* was under way but still alongside the pier.

Jim looked abeam to starboard and studied the ship's position by lining up a light post on Pier 7 with a similar post on Pier 8. Very slight movement aft. "Indicate four knots." The astern movement ceased and a few seconds later he detected slight forward movement. "Indicate three and a half knots." The ship stabilized with very little movement fore or aft. "Time for Phase 2," Jim said in a low voice, more to himself than anyone else. Captain Lynch smiled.

In a loud clear voice Jim said, "Right full rudder. Train Port and Starboard APUs to 225. Indicate four and a half knots." As his orders were acknowledged, Jim looked aft closely to detect movement, then he looked at the bow, then the stern again. *There, the stern's moving out,* he thought, and when he looked forward, *and now the bow is moving. Good.* The ship was moving laterally to port, slowly away from Pier 7, toward the center of the slip. After a while he could see that the stern was moving away from Pier 7 more than the bow. No longer parallel with the pier heading of 053, the ship's heading was now 056. "Indicate three and a half knots," Jim ordered, and a few seconds later the stern movement to port slowed as the ship started to move aft. Looking at the bows of both cruisers as he studied the scene aft, Jim asked the fantail for distance to the bow of the outboard cruiser. "150 feet," came the reply as *McClusky* continued to move toward the center of the slip, diagonally now with slight sternway away from Pier 7, nearly parallel. Soon Jim could see the port side of the outboard cruiser as his ship continued to move laterally.

Jim walked through the pilot house to the port bridge wing, pointed to the ships at Pier 6 and asked the quartermaster holding a laser rangefinder, "Distance to the amphib?" "225 feet," was the reply. And the captain asked, "Jim, what's the distance, skin to skin, when we're centered between that outboard cruiser and the gator?" Jim did not have to look in his wheel book. "190 feet, Sir, almost there," he said. As he and the captain went back to the starboard bridge wing, Jim told the JOOD to stay on the port wing with the QM and, "Let me know when it's 190 feet. I want to keep my eye on the other side." Aft of *McClusky* there was now opening water along the full port side of the outboard cruiser as they moved slowly toward the center of the slip with slight sternway. "Stern's passed the bow of the cruiser," was the report as the JOOD called. "190 feet to the gator." Jim looked at the speed indicator showing minus 0.6 knots. Only Captain Lynch heard as Jim mumbled, "Time for Phase 3." Jim asked the JOOD to bring the QM with the laser rangefinder to the starboard bridge wing and the first distance to the cruiser was 200 feet, drawing a thumbs-up from the captain.

"Engine stop. Rudder amidships. Train port and starboard APUs to 180," Jim ordered, and to the captain he said, "Looks like the amphibs are giving us some windbreak." The speed indicator showed minus 1.1 knots, then 1.4 then 1.9 as they increased sternway, driven by the APUs. "I haven't seen that merchant yet," Jim said to himself, then to the phone talker, "Fantail, bridge, is there any traffic in the channel?" Moments later the talker reported, "Large car-carrier off the starboard quarter, about 1,500 yards, headed outbound."

"That's the *Honda Maru,*" said Jim. Then to the JOOD he said, "Tell him we will wait for him to pass." Jim then ordered "Engine ahead two-thirds. Port and Starboard APUs stop." The ship slowed, and just before it stopped, Jim ordered, "Engine stop." *McClusky* was poised in the center of the slip, while wind and current were setting her down on the cruiser at Pier 7.

"Give me ranges to the cruiser," Jim ordered as they all watched the bow, and then the long blue-and-white hull, pass across their stern, heading toward sea. "175 feet to the cruiser," reported the QM. The big merchant ship hull was almost clear. "150 feet," then "125 feet."

Jim ordered in a clear loud, calm voice, "Right full rudder. Engine ahead one-third, indicate four knots. Start port APU." As the orders were acknowledged, he said, "Start starboard APU. Train Port and Starboard APUs to 225." He looked at Captain Lynch. "Looks like I'm back to Phase 2." The QM called out "100 feet to the cruiser."

"One hundred feet," then "100 feet," and then "115 feet" were the reports as they watched *Honda Maru* clear Pier 6 and continue up the channel. "Rudder amidships. Engine stop. Train Port and Starboard APUs to 180," Jim ordered, and *McClusky* started, once more, to leave the slip. A few seconds later, fantail reported that the stern was clear of the slip and "Channel clear to starboard. Merchant ship in channel opening to port." As the ship left the protected lee of the amphib ship at Pier 6, wind and current started swinging the stern south as fo'c'sle reported, "Bow clear of the end of the pier." Jim mumbled, "Time for Phase 4," as Captain Lynch went to his seat in the pilot house.

"Five hundred yards to shoal water," the navigator called out as Jim looked aft. On the starboard quarter he could see Buoy 30 marking the far side of the channel. He waited until the bow was well clear of the piers and as the speed indicator showed minus 1.7 knots, "Right full rudder" caused the stern to move a little more to starboard. As the navigator called, "Entering the channel, three hundred yards to shoal water astern," Jim ordered, "Engine ahead two-thirds. Shift your rudder. Train port and starboard APUs to 225." *McClusky* shuddered slightly as she rapidly lost sternway, stern swinging to starboard, bow to port, came to zero speed, then gradually gained headway, still twisting in midchannel. Jim walked into the pilot house and stood at the pelorus.

"I hold you in midchannel," called the navigator. "Shoal water two hundred yards to port. Recommend first leg course 328. Seven hundred yards to turning point for next leg." Jim responded, "Very well," then ordered, "Rudder amidships. Stop port and starboard APUs. Train port and starboard APUs to zero zero zero. Retract port and starboard APUs." As his orders were acknowledged, Jim looked at the gyro repeater. "Left 10 degrees rudder. Steady on 328," and *McClusky* started Phase 5, the egress from San Diego Bay via the main shipping channel.

"When Buoy 28 is abeam to port, recommend coming left to 312 for next leg. Next leg three miles, eighteen minutes at ten knots." The navigator's recommendation was acknowledged as *McClusky* continued her harbor transit.

Captain Lynch sipped his coffee and looked at Jim. "Looks like you didn't get that windbreak you expected from those amphibs, Jim," he said.

"No, Sir, but I didn't plan on having to loiter between the piers. I guess I should have gone upwind more, closer to the amphibs, against the wind and current."

"Yes, maybe so, but you had full control with your Phase 2 dance. Nice use of engine, rudder and APUs."

"Thank you, Sir. It was just like in the simulator."

Some of the most interesting and challenging shiphandling is getting under way and landing. The proximity of solid objects demands precision from the conning officer. As with all shiphandling, advance planning is the key to execution. It would be impossible to count the number of times commanding officers have instructed apprentice shiphandlers, "Tell me what you plan to do." This has now been expanded to a formal navigation brief at which a "voyage plan" is presented for each event. Yet as the opening vignette illustrates, the commanding officer is still likely to ask the conning officer for his intentions. For most evolutions there are several different ways to accomplish your purpose, and the number of variables ensures that no example is exactly like another. The key is to understand and use the various forces working on the ship

Knowing Your Ship

Your plan needs to be based upon both general and specific information. General information is that which remains the same. It includes the turning characteristics of your ship, how quickly she accelerates and stops, and how she reacts to wind and current. Specific information is that which changes from one evolution to the next, including wind, current, mooring location, and traffic in the channel.

It is particularly important to have a feel for the lag time between when an order is given and when it takes effect. The engine response time for a ship with variable-pitch propellers is generally faster than for ships with fixed-pitch propellers. When alongside a pier, you usually do not want to wait until the ship starts to move in response to an engine order before ordering a stop.

You also need to know how your ship behaves with opposed engines. While in theory opposed engines should balance, this is often not the case. When you are twisting with one engine ahead one-third and the other engine back one-third, do you gather headway or sternway? This may vary with other combinations. For example, a ship that gathers sternway with a one-third twist, may gather headway with a two-thirds twist, or vice versa. If your ship has auxiliary propulsion units, you need to know what amount of thrust from the main engine is needed to counterbalance the thrust from the APUs at various angles of train. There is no substitute for having experimented with your ship to determine how she behaves under a variety of circumstances.

The greater the wind and current, the more important it is to know how your ship responds to these. If your ship, like many warships, is reluctant to twist into the wind, this must be taken into account in your plan. Otherwise

Figure 4–3. Guided missile destroyers USS *Milius* (DDG 69) and USS *Higgins* (DDG 76) nested alongside the pier in Singapore. *U.S. Navy photo*

you may find yourself being blown sideways toward shoal water, while ineffectively trying to twist your bow up into the wind.

The View from the Bridge

If your ship has been designed with the needs of the shiphandler in mind, the bridge wings will extend far enough that you can see the sides, the jackstaff at the stem, and sometimes even the stern flagstaff. On even the best designed bridge, however, there is a distance forward of the bow and aft of the stern that is out of sight. This becomes increasingly important as we have to place our ship alongside between obstructions, such as other ships, dolphins, or the shore end of the pier. To handle the ship with precision we have to know accurately distances to objects both ahead and astern. Yet there is always an "invisible distance" which is the distance ahead or astern which is obscured by your own ship's structure. You need to determine the invisible distances for your ship. With this knowledge you can determine exactly the distance to an object ahead at the moment it begins to disappear from view. Obviously this will vary somewhat with the height of the observer, but for all practical purposes an average height of eye can be assumed (see fig. 4–4).

It is helpful to have an experienced person on both forward and after line-handling stations send up estimates of distance to objects of interest. With the advent of laser rangefinders these reports have become more accurate. Still, no shiphandler prefers to be completely dependent upon relayed information. As with virtually everything having to do with shiphandling, it is good to have

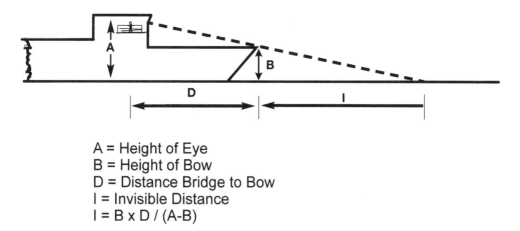

A = Height of Eye
B = Height of Bow
D = Distance Bridge to Bow
I = Invisible Distance
I = B x D / (A-B)

Figure 4–4. Invisible distance.

multiple sources of information. Many ships place marks on the jackstaff that provide range estimation when approaching objects ahead. These marks are placed so that when standing in the normal conning position the line of sight through the marks to the waterline of an object ahead corresponds to preselected ranges, such as 1,000 yards, 500 yards, 250 yards, and so on.

Visibility astern is usually considerably worse than ahead, but it is vital to know when the stern is free of an obstruction. To develop a visual reference, when circumstances permit place an object in the water lined up laterally with the stern. Then from the conning position on the bridge wing select a fixed point on the ship that lines up with the object. It can be useful to paint a reference mark at this point. Once this reference is determined, any obstruction astern whose waterline is above the reference point is clear of the stern.

Evolution-Specific Information

Specific local information you can collect ahead of time includes the heading of the pier at your assigned berth, depths of water and any navigational hazards along your intended track, whether the pier is pilings or solid face, and the predicted wind and current. Your plan will be affected by the ship's position alongside the pier. Are you headed bow out or bow in? Alongside the pier or tied up alongside another ship? At the end of the pier or toward the head? What other ships are moored in locations that can affect your plan? Once under way, what maneuver must you perform to wind up headed fair in the channel? All of these considerations and more must be taken into account in your plan for getting under way.

Once you have made your plan it is a good idea to brief the line handler supervisors of your intentions, if this has not already been done at the planning conference. If everyone knows what to expect, the likelihood of a smooth and professional operation is greatly increased. It is also always a good idea to make sure of your communications with the line-handling parties well in advance, to include talkers thoroughly familiar with the line-handling standard commands to be expected.

No two circumstances are ever exactly the same, but if you understand the principles involved they may be used in almost any situation that arises. The first rule of planning to get under way or make a landing with a ship is to leave yourself a margin. If you plan a maneuver that requires full rudder and large backing bells, there is little you can do if you have miscalculated, if response is not instantaneous, or if there is a material casualty. Murphy's

Law, which states that if something can go wrong, it will go wrong, applies to shiphandling as to other things in life. If, on the other hand, you have planned your maneuver so that it works with less rudder and smaller bells, you have a much better chance to recover if you have miscalculated or if Murphy puts in an appearance. There are times and places where operational necessity or extremes of wind or weather require a shiphandler to run a calculated risk. These times are rare, and there is no excuse for risking damage to your ship in an attempt to appear dashing.

We will concentrate in this chapter on evolutions to be carried out by the ship alone without the assistance of tugs. The larger the ship or the more difficult the evolution, the more likely tugs will be involved. The use of tugs is discussed in chapter 8. The discussion here proceeds in a rough order of degree of difficulty, from getting under way with no wind or current to somewhat more difficult situations, for both single- and twin-screw ships. To simplify the discussion the term "pier" is used generically to refer to any shore structure to which the ship is moored. More precise definitions are set forth below:

Bitt: Pair of short steel posts or horns on board ship used to secure lines.

Bollard: Steel or iron post on a dock, pier, or wharf used in securing a ship's lines.

Bullnose: Closed chock at the bow of a vessel. Has the appearance of a large flared nostril.

Camel: Float used as fender between two ships or a ship and a pier. Also called breasting float.

Chock: Metal fitting through which hawsers and lines are passed. May be open or closed. Also blocks used to prevent aircraft or vehicles from rolling or blocks used to support a boat under repair.

Cleat: An anvil-shaped fitting for securing or belaying lines.

Dock: Large basin either permanently filled with water (wet dock) or capable of being filled and drained (drydock or graving dock).

Dolphin: A cluster or clump of piles used for mooring. A single pile or a bollard on a pier is sometimes called a dolphin.

Pier: Structure for mooring vessels which is built out into the water perpendicular to the shore line.

Piling: Wooden, concrete, or metal poles driven into the river or sea bottom for support or protection of piers or wharves.

Quay: A solid stone or masonry structure built along the shore of a harbor to which boats and ships make fast, load, unload, etc.

Slip: A narrow stretch of water between two piers.

Wharf: Structure parallel to the shoreline to which ships moor for loading, unloading or repairs. Sometimes called a quay, which is usually a solid masonry structure.[1]

Getting Under Way

Your plan for getting under way must encompass more than just getting clear from alongside. Only rarely is it possible to move directly from alongside to headed fair in the channel. More normally you must thread your way past other piers and other ships. Frequently a sharp turn is necessary as you clear the pier and turn into the channel. Shoal water may be close along the way. Thus your plan must include the sequence of maneuvers that will move your ship smoothly from the pier into the channel. You also need to anticipate the "what ifs." What if traffic in the channel interferes? Will wind or current present a problem if you have to wait? What would you do if an engineering casualty caused a loss of power or steering? With luck, none of these things will happen. But if they do, you will be better prepared for having thought them through in advance.

No Wind or Current

Dead calm conditions are rarely encountered in the real world but are a good place to start gaining an understanding of how to get a ship under way from alongside a pier or another ship. Getting under way is usually easier than making a landing, if only because in the course of the maneuver you get progressively further away from danger in the shape of the pier or the ship alongside. The first consideration is whether you will be getting under way bow or stern first. If stern first, and there is no wind, current, or obstruction, this is a relatively simple evolution of getting the ship's stern out far enough to be able to back clear.

If yours is a single-screw ship, make sure to anticipate the stern walk you will experience when backing. The stern will walk to port when backing for most single-screw ships. The *Oliver Hazard Perry*–class frigates are an exception, in that they experience starboard sternwalk under almost all circumstances, even when at zero pitch.

There are several ways of moving the stern away from the pier, most of which involve moving the bow toward the pier. Make sure properly rigged fenders are in place at the point at which the forward part of the ship will contact the pier or camels. If your ship has a large bow mounted sonar dome caution is necessary to avoid bringing the bow too close to the pier while getting the stern out. A good camel between the ship and pier makes this easier. If you are alongside another ship, or if there are ships moored aft, in addition to fenders make sure the anchor on the side toward the pier is dipped.

After singling up, try slacking all lines. This will give you a feel for how the ship will move when the lines are taken in. Some of the ways of moving the stern out follow:

1. Slack or take in all lines except the bow breast line (line number one) and the forward after spring line (line number two) (see fig. 4–5). Using the capstan, heave around on number one to bring the bow in to the pier, moving the stern away from the pier. If this brings the stern out far enough to be well clear of any obstructions astern, take in all lines and back clear. If this does not bring the stern out far enough, hold line two and go ahead gently on the outboard engine with the rudder turned toward the pier. This same thing can be done with a single-screw ship, going ahead gently with the rudder turned toward the pier and holding line number two to keep the ship from moving forward.

2. With a twin-screw ship, the stern can be twisted out by backing one-third on the engine next to the pier and going ahead one-third on the outboard engine, with the rudder full toward the pier. It is important not to get way on the ship before you are properly positioned. It is a good idea to keep lines one and two available until you are satisfied with the ship's alignment

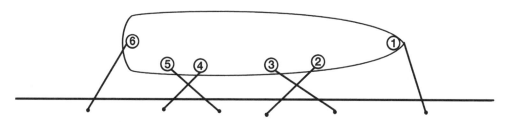

1. Bow Line
2. After Bow Spring Line
3. Forward Bow Spring Line
4. After Quarter Spring Line
5. Forward Quarter Spring Line
6. Stern Line

Figure 4–5. Mooring lines.

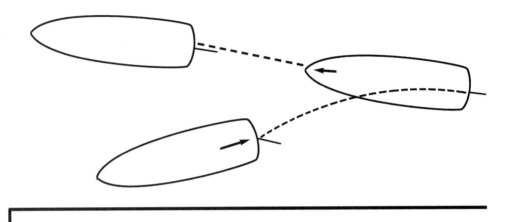

Figure 4–6. Going out ahead.

for backing clear. If the pier is solid, the wash from the backing engine will help to move the ship away from the pier.

Going out ahead is somewhat more difficult, if only because the propellers turning ahead draw water from between your own ship and the ship or pier alongside. This means that you need more separation before going out ahead. You can try taking line six to the capstan while slacking lines one, two, and three, to see if this can get you enough of an angle to go out ahead. If room astern permits, one way to get clearance from the pier or ship alongside is to get the stern out, as described above, then back away with the rudder full toward the pier to swing the bow away from the pier (see fig. 4–6). When the bow is pointed sufficiently away from the pier you can then go out ahead. As soon as your ship's pivot point passes the end of the pier or the bow of the ship alongside, rudder can be used to swing the stern out for additional clearance, if needed.

An Offsetting Wind or Current

It is simple to get under way with an offsetting wind and or current. After singling up all lines, if going out astern slack the after lines to bring the stern out to the desired angle, then take in all lines and back clear. The potential hazard here is that as you back out of the slip your stern will come into the full effect of the current before the forward part of the ship. If not anticipated, this can create a strong rotating force on the ship, swinging the stern away from the pier, and thus moving the bow toward the pier. The stronger the current, the more power should be used in backing clear of the slip, in order to minimize the length of time the ship is subjected to this rotation. If going out ahead, slack the forward lines until the bow is out sufficiently to take in all lines and go out ahead. In this case, the bow will enter the stronger current first, and the resulting force rotating the stern toward the pier needs to be taken into account. As before, once the pivot point is past the end of the pier or ship alongside, rudder may be used as needed to swing the stern away from the pier.

An Onsetting Wind or Current

One of the most difficult circumstances for getting under way is when a strong wind or current is holding the ship to the pier. In this situation, the means discussed above for moving the stern out may be used, but it is necessary to get a larger angle with the pier before starting to back, since the ship will be set to leeward. It is also desirable to use substantial power, generally all back two-thirds, or in the case of an exceptionally strong current even a back full bell may be indicated. If a safe angle with the pier cannot be achieved prior to taking in your lines, it may be time to wait for a lessening of the current, or call for the services of a tug.

Going out ahead is even more difficult under these circumstances, and most times she will require the services of a tug. If an onsetting wind or current can be anticipated, an anchor can be dropped during the approach to the pier and used later as an aid to moving the bow away from the pier. See a discussion of this in chapter 6.

Wind or Current from Astern

If getting under way stern first with the wind or current from astern, bring the bow in using the bow breast line (number one) to the winch. Hold the after spring lines (numbers two and four) to keep the ship from being moved ahead by the current. Slack the stern breast line (number six) enough to let the current move the stern out to the desired angle, then take in all lines and back clear. It is important not to lose control of the stern, since once the current is on the inboard side it will be trying to rotate the ship. Be ready to take in lines and get under way briskly when the desired angle is achieved.

To go out ahead with a current from astern, hold the after spring lines (numbers two and four) to keep the ship from moving ahead. Take in three and five. Slack the stern line (number six) enough for the current to start moving the ship away from the pier, then tend one and six to allow the ship to sail sideways away from the pier. If the stern begins to come out too far, a back bell into the current will help the alignment. When separation is sufficient, take in all lines and go out ahead. Remember that speed over the ground will be the sum of the current and the ship's own speed through the water.

Wind or Current from Ahead

Wind or current from ahead can be used to move the ship laterally away from the pier. The technique involved is to slack forward lines enough to get the current on the inboard bow, checking after spring lines enough to prevent the current from moving the ship aft. With the current on the inboard bow, the ship will sail away from the pier. When enough lateral distance is obtained, to go out stern first, go ahead with rudder toward the pier to swing the stern out, then back clear. Anticipate that the speed of the current will be added to the ship's sternway through the water. Steering control while backing with the current is not as predictable as we would like, so give yourself an extra margin when executing this maneuver.

To go out ahead with current from ahead, slack lines one, two, and three to allow the bow to come out under control. With the current on the inboard side, slack all lines to allow the ship to move laterally from the pier, with the bow coming out further. Come ahead on engines as needed to counteract the current. When alignment and separation are adequate, come out ahead.

5

MAKING A LANDING

◄◦►

"**B**uoy 26 abeam to starboard. Hold you midchannel, on track," the navigator called out. Ens. Ed Moore replied with the customary "Very well." Ed was conning officer of USS *Cowpens* (CG 63), and as such, he had directed the movements of his ship from before they had reached the sea buoy, 1 SD, to this point, passing under the Coronado Bridge and approaching Naval Station, San Diego. It was Friday afternoon and *Cowpens* was returning to its home port, having spent three days exercising at sea in the Southern California Operating Areas.

Ed exchanged nods with Lt. John Williams, who had been standing at the rear of the pilot house, as John walked forward, saluted, and said, "I am ready to relieve you, Sir." Ed returned the salute. "I am ready to be relieved, Sir. We're steering 132, making ten knots, 55 rpm with 95 percent pitch both engines. No traffic. Just passed Buoy 26." Ed nodded toward the starboard quarter. "En route to Pier 7, just coming in sight." He pointed ahead. "You can see the bow of an LSD at the south side of Pier 6 and the stern of *Shiloh* at Pier 7." John raised his binoculars to his eyes. "Got it." He lowered his binoculars and saluted. "I relieve you, Sir." Ed returned the salute and replied, "I stand relieved, Sir."

Captain Bethea's shiphandling program gave the junior officers every opportunity to handle the ship. Today was Ens. Ed Moore's second chance at conning *Cowpens* into port, and as planned, as they reached the naval station, a more experienced officer took the conn to make the landing. Lt. John Williams was "SWO," a qualified surface warfare officer. He had made landings before in his previous ship, but

Figure 5–1. USS *Shiloh* (CG67) heads into her berth at Naval Station San Diego at the end of a ten-month deployment to the Arabian Gulf. *U.S. Navy photo*

this would be his first in *Cowpens*. He had been watching at the rear of the pilot house since they entered the channel.

"This is Lieutenant Williams, I have the conn," John announced.

The helmsman acknowledged with "Aye aye, Sir, steering 132, checking 119 magnetic," followed by the lee helmsman: "All engines ahead two-thirds, turns and pitch for ten knots."

John turned and saluted the captain, who was seated on the starboard side of the pilot house, binoculars to his eyes. "Captain, I have the conn, Sir." Captain Bethea replied, "Very well," lowered his binoculars, returned the salute, and asked, "How are you going to make this landing, John? Let's go over it again."

"Sir, the things we're concerned about are wind, current, depth of water, turning space for approach, length of pier space for us, and clearance between ships at the south side of Pier 6 and north side of Pier 7." The captain nodded as John continued, using his flat left hand as a model ship. "Wind is ten knots from 120 and we'll have an ebb current of one-half knot. Both will be setting us off the pier." John's right index and middle fingers demonstrated the external forces as Captain Bethea asked, "And what is the total effect, in equivalent wind, of the half knot of current with the actual ten knots of wind?"

"Well, one knot of current is roughly equivalent to thirty knots of wind, so we'll have about fifteen knots of wind effect caused by current, and with ten knots of actual wind, that will be equivalent to twenty-five knots of wind setting us off," John calculated aloud. "But once we start into the slip, the ships at the piers should break the wind, some, and current will be less closer to the quay wall. So I expect an equivalent of twenty-five knots of wind setting us northwest or off the pier as we approach the slip"—John again demonstrated with his hands—"about eighteen knots as we pass *Shiloh* and *Antietam,* then about twenty knots as we go into our berth at Pier 7."

"Okay, and do you think we can handle that offset with only one tug helping?"

"Yes, Sir. With one tug and our ship's power we'll be fine."

"How do you know?" Captain Bethea asked, smiling.

"Well, Sir, I haven't actually done this before, in this ship and with these conditions, but I have done it a number of times in the shiphandling simulators at Newport and Norfolk, with this class ship and under these and even worse conditions. I can do it, Captain."

"I'm sure you can." The captain then asked, "How about water depth?"

John explained the charted water depth with state of tide and compared *Cowpens* draft with predicted depth, then added that frequent *Ticonderoga*-class cruiser use and deeper draft ship use proved water depth to be more than adequate.

"There's about five hundred yards of good water from the pier head across the channel to shoal water, so if we hug the right side of the channel we'll have plenty of room to turn-in and get aligned before we enter the slip."

"Okay," the captain replied, "but don't forget that when you turn your stern will be swinging toward the shoal water to starboard, so you need to make allowances."

"Yes, Sir," said John. "I've taken that into account." He was looking in his small pocket notebook as he continued. "Pier 7 is 1,450 feet long, and *Antietam* occupies 567 feet of that plus 20 feet from its stern to the pier end, so that's 587 feet of pier space, leaving 863 feet for us. Our length of 567 leaves us with 148 feet from our stern to the bow of *Antietam*"—John pointed aft—"and 148 feet from our bow to the quay wall." Jim pointed forward as *Cowpens* proceeded in the channel past Pier 3.

"And how about clearance between ships, between the amphibs at Pier 6 and the cruisers at Pier 7?" The captain pointed left and right. (See fig. 5–2.)

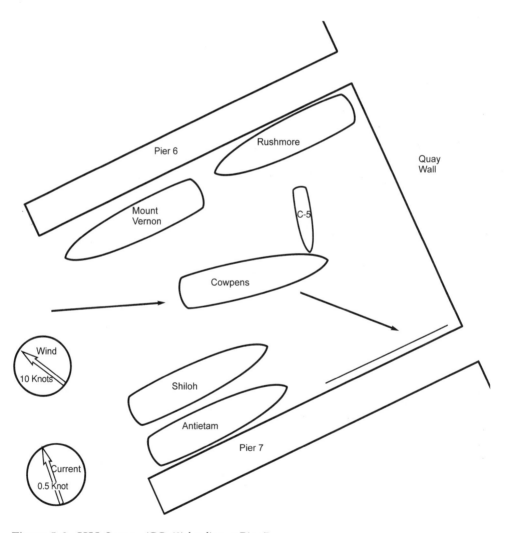

Figure 5–2. USS *Cowpens* (CG 63) landing at Pier 7.

John looked in his notebook again. "There's 650 feet of water between the piers," he replied. "Subtracting the beams of two cruisers and an LSD, and their fenders, leaves 438 feet of water for us. That means 190 feet skin-to-skin to the portside LSD and 190 feet to starboard, to the outboard cruiser *Shiloh,* for us to be in the middle of the opening. Plenty of water."

"Yes, plenty of water, as long as that wind and current don't set us down on those amphibs." The captain smiled. "Go ahead, John, make the landing."

"Indicate maneuvering bells. All engines ahead one-third," John ordered, and lee helm responded.

"Hold you midchannel on track, approaching Buoy 28 to starboard," the navigator called out. "With Buoy 28 abeam to starboard recommend coming right to channel course 148 for final leg."

"Conn, aye," John responded. "I'm going to turn early to set up on the right side of the channel. It'll help with the turn into the slip." "Navigator concurs," was the response, and a few seconds later John ordered, "Right full rudder. Steady on 148." The helmsman acknowledged, and *Cowpens* turned to the new course.

"Charlie 5, this is warship 63," John said. "I'm slowing now. I'd like you to make-up with a headline on my port bow." The tractor tugboat had been following since *Cowpens* had passed under the bridge, waiting for the ship to slow and for positioning orders. "Charlie 5, roger, out," came the snappy reply.

With Pier 5 abeam to port, John ordered, "Left full rudder. Port engine stop." Helm and lee helm acknowledged as the tug control radio circuit crackled, "Charlie 5 made-up on your port bow. Ready to work." John acknowledged as his ship slowed, swinging to port, toward Pier 6 with the bow of *Mount Vernon* on the port bow of *Cowpens.* "Charlie 5, bow to port, easy," put the tug to work helping swing *Cowpens* into the slip.

Helm had been calling out headings every ten degrees since John had ordered the turn: "Passing 130. Passing 120. Passing 110." As the ship's heading passed 070, John ordered, "Rudder amidships." Helm responded as John told the tug, "Charlie 5, stop," and *Cowpens'* swing to port slowed nearing the pier heading of 053.

John was standing just aft of the centerline pelorus in the pilot house as he viewed his alignment and spacing. On the port bow was LSD *Mount Vernon* at Pier 6, and on the starboard bow was cruiser

Shiloh, outboard of *Antietam* at Pier 7. He wanted his ship to be positioned equally between the LSD and *Shiloh* as he entered the slip, and on a heading parallel to the pier, but John could see that the wind and current were pushing him toward the LSD and he had swung past 053. John announced, "Looks like we're being set to the left," and the captain agreed, adding, "Just what we expected."

"Charlie 5, bow to starboard, easy," John ordered. He received a terse "Bow starboard, easy" from the tug. Then he ordered, "Left full rudder," and helm responded. A few seconds later John ordered, "Port engine back one-third," and lee helm acknowledged with "Port engine back one-third. Starboard engine ahead one-third." John carefully watched the movement of his ship.

John turned to the quartermaster holding a laser rangefinder. Pointing at *Mount Vernon*'s bow, he directed, "As soon as we get into the slip, give me distances to that amphib."

The bridge phone talker called out, "Fo'c'sle reports bow crossing the pier head, entering the slip," and John acknowledged with, "Conn, aye." He could see that *Cowpens* was still being set to port, toward the LSD. "Starboard engine ahead two-thirds," he ordered. And as lee helm was acknowledging, he ordered, "Charlie 5, bow to starboard, half" on the tug control circuit. "Bow starboard, half," came the response. As John walked out on the starboard bridge wing, he could detect movement to the right as his ship proceeded past the cruiser.

"One hundred fifty feet to the amphib on the port side," the quartermaster called out from the other bridge wing. John looked at his starboard side spacing to *Shiloh.* "More than that here," he said. "We've been set about fifty feet." The captain had moved to the starboard bridge wing and said, "Looks like you caught it with the two-third bell." "Yes, Sir. I've got to slow now."

"Port engine back two-thirds. Charlie 5, bow to starboard easy." Distance to the cruiser was reduced and *Cowpens* was parallel to the pier. The quartermaster called out, "160 feet to the amphib," then "170 feet." A few seconds later, "180 feet," as *Cowpens*'s bridge glided slowly past the bridge of *Shiloh.*

"Starboard engine ahead one-third," caused a noticeable slowing as *Cowpens*'s bridge passed the bow of *Shiloh.* "One hundred ninety feet to the amphib" brought smiles to captain and conning officer. "Now we'll see if we get more wind effect as we clear the cruisers," cautioned the captain.

But the wind and current effect setting them off the pier did not change noticeably, and *Cowpens* slowed, almost to a stop, creeping forward at half a knot as John ordered, "Port engine back one-third."

"Fantail reports stern clear of the cruiser," sang out the phone talker as John acknowledged and carefully watched his slow forward movement with slight movement against the wind and current, toward the pier. The ship seemed to be hovering, with tug, engines, and rudder holding position against wind and current. "Looks like we'll have to use more power to get alongside the pier," John mumbled. The captain nodded as John ordered, "Starboard engine ahead two-thirds. Port engine back two-thirds" and "Charlie 5, bow to the pier half." There was a slight shudder and *Cowpens* moved aft and sideways, toward the pier, bow slightly in. "Port engine back one-third" stopped the movement aft, and "Charlie 5, bow to the pier easy" brought the ship more parallel to the pier. "Starboard engine ahead one-third," John ordered.

The quartermaster called out, "380 feet from the amphib," and John acknowledged and ordered, "Belay your distances to the amphib." They continued to close the pier as John turned to the phone talker and said, "Fo'c'sle, fantail, bridge, over all lines." Soon heaving lines were flying out to the pier, where line handlers quickly pulled mooring lines to bollards. The Bridge Here sign on the pier platform was forward of the bridge. He had to move the ship forward.

"Fo'c'sle reports lines 1, 2, and 3 to the pier and slack. Fantail reports lines 4, 5, and 6 to the pier and slack," called out the phone talker. John responded, "Bridge, aye. Tell Fo'c'sle and fantail to keep the slack out of all lines. We've got to move about thirty feet forward."

"Starboard engine ahead two-thirds," John ordered, and as soon as that order was acknowledged, "Starboard engine ahead one-third." *Cowpens* moved slightly forward, hugging the pier, the Bridge Here sign almost at the bridge wing. Not enough. John repeated the "touch" of the starboard engine to two-thirds then immediately returned to one-third power. A few feet forward was sufficient to match the Bridge Here sign. The ship hovered against the offsetting wind and current, as before, but this time alongside Pier 7 in proper mooring position. John ordered, "Hold all lines," and when that had been accomplished, "Charlie 5, stop," and then, "All engines stop. Rudder amidships." To John's look, the captain nodded and said, "Double-up." John turned to the phone talker: "Fo'c'sle, Fantail, Bridge. Double-up all lines."

A few minutes later, the phone talker reported all lines doubled, and John picked up the handset for the tug control circuit. "Charlie 5, finished. Thank you for your service. Take in your line." As he was telling fo'c'sle to cast off the tug, he heard, "Roger. A pleasure to be of service. Out." To the phone talker he said, "CCS, bridge, finished with main engines."

The bos'un trilled steady-up-steady-up sharp followed by "Secure the special sea and anchor detail with the exception of line handlers. Set the regular in-port watch. On deck section two." The bridge/pilot house team started putting away phones and navigation equipment. Captain Bethea and Lt. John Williams remained on the starboard bridge wing discussing the landing, a left flat hand serving as ship model, right index and middle fingers the wind and current.

Landing Alongside a Ship or Pier

Advance planning of the maneuver is essential to doing a good job of landing a ship alongside. The plan must include accounting for the assigned berth, the velocity and direction of the wind and current, whether you are going pierside or alongside another ship, whether the pier is piling or solid face, the positioning of nearby ships, potential traffic in the channel, and so on.

A key part of any landing is setting up for your approach. This is where the most mistakes are made. A proper plan entails knowing where you want to position your ship alongside and how best to achieve a smooth and controlled transition from the channel to your assigned berth. If possible, your approach should be lined up from a good distance out, rather than relying on a tight turn into position. As you slow for your approach the ship becomes less responsive to its rudder, but the effect of wind and current continues undiminished.

Accurate anticipation of the effects of wind and current are essential. By all means calculate the tidal current and read the anemometer, but do not rely exclusively on instrumentation. Get out on the wing of the bridge to feel the wind on your face, and observe it on flags and the surface of the water. Is it steady or gusting? From the anticipated direction? Read the current off of buoys and pilings. Does its direction and velocity appear to accord with what was calculated? A good shiphandler is continuously aware of wind, weather, and current.

The effect of wind and current on your landing will vary greatly depending not only on their velocity, but on their direction. Wind or current on the ship's beam will have much more effect than on the bow or stern. A useful rule of thumb as a first approximation is that one knot of current is equal to thirty knots of wind, but you need to learn how your own ship reacts and be aware of how it changes with conditions of loading. The more hull below the waterline, the greater the effect of current. The more sail area above the waterline, the greater the effect of wind. For example, an *Arleigh Burke*–class DDG will be more sensitive to current than a *Ticonderoga*–class CG. In comparison, the CG, with more sail area, is more affected by the wind. All other factors being equal, the lighter the load, the greater the effect of the wind.

It is easy to make the error of bringing your ship in too close to the pier. So long as your lines will reach, they can be used to move the ship alongside. In general, the larger the ship, the wider the approach should be and the more the ship should be kept parallel to the pier. Large ships will use tugs for most evolutions. The use of tugs is described in more detail in chapter 8.

As you transition from the channel to set up an approach to your assigned berth, you will slow the ship. Once you slow, give rudder orders, not courses for the helmsman to steer. Remember that at slow speeds the rudder is relatively ineffective unless an ahead bell is discharging over it. Once the approach has begun, it is the responsibility of the conning officer, not the helmsman, to decide where the rudder should be.

Landing with No Wind or Current

The simplest landing is one in which there is no significant current and our assigned berth is at the seaward end of the pier. Such benign conditions are rare but are a good place to start. With a single-screw ship that backs to port, the easiest landing is port side to. The opposite is true for a single-screw ship with a variable-pitch propeller, such as the *Oliver Hazard Perry* class of guided-missile frigates. The *Perry* screw rotates clockwise whether going ahead, astern, or stopped. Thus the stern tends to walk to starboard under all circumstances. The extra maneuverability provided by the APUs, however, makes the *Perry* class one of the most maneuverable of warships. (See the special section on the *Perry* class in chapter 13.) For the remainder of this chapter, when reference is made to a single-screw ship, it applies to the larger variety of ships that back by reversing the direction of shaft rotation.

In the absence of significant current, either a twin-screw ship or a single-screw ship going port side to should make their approach at an angle of 15–20 degrees to the pier heading. The bow should be aimed at a point about ten to fifteen yards outboard of the intended final position of the bridge. With no current, the approach can be made slowly, stopping engines to allow the ship to coast into position, using standard rudder away from the pier to swing the stern in as the bow lines go over to the pier. The goal is to wind up in a position stopped parallel to the pier about ten yards out, then use mooring lines and winches to move in to the pier. If no winch is available aft, after the bow lines are in position a twin-screw ship can be twisted in by going ahead one-third on the pierside engine and back one-third on the outboard engine. With a single-screw ship port side to, the stern can be moved in by holding the after spring (number two) and going ahead one-third with the rudder to starboard. This should swing the stern in neatly, but be cautious not to leave the bell on too long or to get excessive strain on your spring line. Alternating short ahead and astern bells with the rudder at right full will help move the stern to port. Since you will not be permitting the ship to gather significant way, there is no need to shift the rudder. The ahead bell will thrust against the rudder, while the back bell will walk the stern to port, with negligible effect on the rudder.

When making a landing starboard side to, the maneuver is the same for a twin-screw ship. For a single-screw ship, however, the tendency of the ship to back to port makes the maneuver somewhat more difficult. Again assuming no significant wind or current and a berth at the end of the pier, the approach of a single-screw ship should be shallower, perhaps a 10-degree angle to the pier. A stop bell should be rung up calculated to let the ship coast almost to a complete stop as it arrives at the assigned position. This is to minimize the need for a back bell, which will walk the stern away from the pier. As the ship nears the assigned position, the rudder should be placed left full to swing the stern toward the pier. A short ahead bell may be needed to start the swing, followed, if needed, by a back bell, which will stop the stern's swing toward the pier. If done right, you now have lines over both forward and aft and can use winches to move the ship laterally in to the pier. If the stern needs to be moved in further, you may be able to do this by using brief ahead bells against a full left rudder, and holding line two to prevent gathering headway. Backing bells need to be held to a minimum, since they will move the stern away from the pier. An alternative way to make a starboard side landing is to use the offside anchor at short stay, as discussed in chapter 6.

Landing While Being Set Off the Pier

If the ship is being set off the pier, a wider approach angle is needed to make sure of getting the forward lines over. The approach should be somewhat faster and closer than in the no wind or current situation, to make sure forward lines are over before we are blown away from the pier. But do not overdo it. Once the lines are over a twin-screw ship or single-screw ship, making a port side landing can move the stern in as discussed above.

If you are unsuccessful in getting lines over before being set away from the pier, there is little to do to salvage the situation other than backing clear and starting a new approach. With a strong offsetting wind a single-screw ship lacking a bow thruster or APU and going starboard side to will need either to use a tug or use the anchor at short stay. With somewhat stronger winds and/or current the twin-screw ship may also need to make use of the tug or anchor.

If the pier is solid faced, the current will lessen as the ship heads into the slip. This means the stern will still have the full force of the current but the bow will be shielded. The forces placed on the ship will rotate the ship. If not anticipated and planned for, this has the potential to create a hazardous situation. A similar effect can exist if a large structure on the pier shields the ship from the wind as she enters the slip.

Landing While Being Set On the Pier

A gentle set onto the pier can be an aid to going alongside. All the conning officer has to do is place the ship parallel to the pier a bit wider than normal, and let the current move the ship alongside the pier. As the current increases, however, the landing becomes more difficult. A miscalculation on the part of the conning officer may find the ship in contact with the pier before reaching the assigned berth. If there are no ships in the way, it should be possible after landing to move ahead to the assigned berth. Make sure to inspect the pier face or the camels before moving alongside. A greater hazard exists if the ship develops enough leeway in being set down on the pier to cause damage. Judgment comes into play here. A relatively small ship going alongside a strong pier with a solid line of camels can accept considerably greater leeway than can a ship of greater tonnage going alongside a less auspicious pier.

Once committed to a landing while being set on, it is vital to have the ship as parallel as possible and with all way off at the time you touch. If, for example, the stern lands first, the ship will rotate, slamming the bow into the pier and endangering a large sonar dome. A twin-screw ship being set onto the pier can twist as needed to keep parallel. A single-screw ship can use short ahead bells against a full rudder to accomplish the same purpose. The single-screw ship must take into account the tendency of a back bell to walk the stern to port while keeping parallel. If the pier has a solid face, a short back bell just before touching can provide a useful water cushion between the ship and the pier. If the beam wind or current is still too much for a safe landing, the alternatives are the use of tugs or the use of the anchor to keep the bow under control.

Landing with Current from Ahead

Landing with current from ahead is relatively easy, since the ship's speed relative to the pier is reduced by the velocity of the current. This permits the ship to use more speed, and thus better rudder control. The approach should be made from a fairly flat angle, perhaps ten degrees. The hazard with a strong current is that if the current gets solidly on the outboard bow it can push the bow in toward the pier too rapidly. Once the forward lines are over, the forward bow spring (line number three) can be used to keep the ship from being pushed aft by the current, and the use of the rudder in the current can be used to bring the stern in at a controlled rate. In an exceptionally strong current from ahead, it may be well to go somewhat past the assigned berth, drop the anchor, then veer chain as needed to drop back into position.

Landing with Current from Astern

A landing with a fair current (one from astern) is a difficult and potentially hazardous maneuver. Circumstances permitting, it is preferable to turn the ship and make an up current approach as discussed above. One way of turning the ship is to use the anchor. Before reaching the assigned berth a turn should be initiated toward it and the upstream anchor dropped. The ship will then pivot on the anchor to head upstream. Engines can be used during the swing to control distance from the pier. Once the ship is headed upstream, the landing can be made as discussed above.

    If there is not room to turn, tugs are not available, and the landing must be made headed downstream, the approach should be made as slowly as consistent with maintaining steerageway. It is important to get the stern lines over quickly, so the approach should be almost flat and close to the pier. Once the lines are over, the engines should be backed to stop the ship's forward motion as the ship swings in to the pier on the after quarter spring (line 4). There will be substantial strain on this line, so it needs to be monitored closely. If the pier is solid faced, a short sharp back bell (two-thirds) will help to put a cushion of water between the ship and pier just before she touches. Do not secure your engines until all lines are doubled securely.

Landing Stern First

A stern-first landing is more difficult than with the bow first, primarily because a ship does not steer as well going astern and is less directionally stable. When backing, the pivot point moves toward the stern, giving the rudder very little leverage with which to work. If you have a long distance to back, you either have to have sufficient sternway to steer (typically five knots or more) or depend upon twisting with engines to maintain directional control. It is generally not a good idea to attempt a stern-first landing in tight quarters without tug assistance. If necessary, however, this is a particularly good time to make use of your anchor to stabilize the bow while controlling the position of the stern with engines.

If room permits, it is possible to use the anchor in making a 270-degree approach to a bow-out landing. Figure 5–3 illustrates this maneuver. It begins by passing the end of the assigned pier and letting go the anchor that will be on the side away from the pier to short stay. The ship then pivots around the anchor with full rudder and a one-third ahead bell, veering chain as necessary to move alongside the pier. The anchor moves the pivot point toward the bow and provides excellent control. Once alongside, the anchor chain is slacked to the bottom and can serve later as an aid in moving the ship away from the pier when getting under way.

Some Observations

To become a good shiphandler around the pier it is necessary to learn to read accurately the ship's motion, both fore and aft and laterally. With experience this becomes instinctive. Before that instinct is developed, however, it is useful to select improvised ranges, such as light posts on one pier lining up with those on another pier. Visual relative motion between such ranges gives a

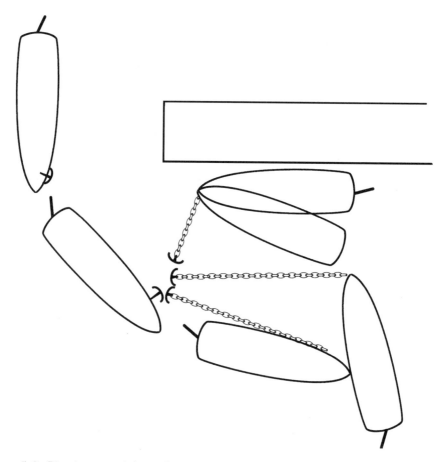

Figure 5–3. Pivoting around the anchor.

very accurate picture of own ship's motion. Use your ship's instrumentation, but do not rely exclusively upon it. Do not forget that, for example, the pitometer log is reading speed relative to the water, not to the solid objects around you. GPS can be a help. But no instrumentation can take the place of your own observation.

When your ship's engines are backing, the water disturbed by the backing screws appears as a swirling pattern on the surface. This pattern of disturbed water is called "quickwater" and can provide a readily observable guide to the ship's velocity through the water (see fig. 5–4). At about three knots, the quickwater moves with the ship. At two knots, it begins to move ahead on either side of the stern. As the backing engines bring the ship to a dead stop, the quickwater moves forward to amidships.

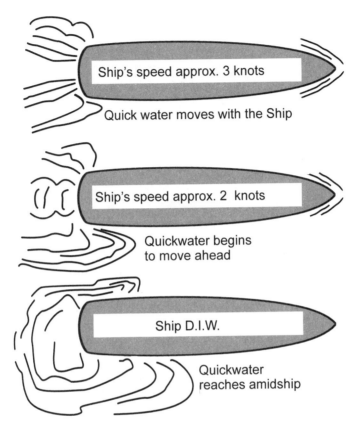

Ship's speed approx. 3 knots

Quick water moves with the Ship

Ship's speed approx. 2 knots

Quickwater begins
to move ahead

Ship D.I.W.

Quickwater
reaches amidship

Figure 5–4. Estimating headway when backing.

It is always a good idea to check that the rudder and the engine order telegraph have moved in the ordered direction. Alongside the pier it is particularly important, for there is less margin to recover from a misunderstood order. Try to anticipate what might go wrong: loss of steering control on the bridge, an engineering casualty, a parted line, and so on, and think through how you would respond. In going alongside, a good plan, well carried out, results in the kind of seamanlike evolution in which the entire ship's company can take pride.

6

GROUND TACKLE

———◄○►———

USS *John Paul Jones* (DDG 53) was at sea off San Diego, California. Lt. (jg) Kathy James was on the fo'c'sle having a conversation with Chief Boatswain Mate Vernon Brown.

"Chief, thank you for taking the time to meet with me. I have been told that I will have the conn for our precision anchoring drill day after tomorrow, and I thought I'd better refresh what I know about what goes on here. I know we'll have a planning meeting for everyone involved, but if I have any dumb questions to ask, I'd rather ask them here."

"Glad to be of help, Ma'am. One thing I've learned after nearly twenty years at sea is that the more people know, the less shouting there seems to be."

"Okay, let's start by your walking me through what I need to know about our ground tackle."

"Yes, Ma'am. *John Paul Jones,* like the rest of the *Arleigh Burke*s in the fleet, has two anchors. Our main anchor is a centerline stockless weighing nine thousand pounds, with 180 fathoms of chain. We have a Mark 2 lightweight Danforth anchor to port. It weighs four thousand pounds and has 120 fathoms of chain."

"I guess that means that we want to use the centerline anchor anytime we need holding power, right?

"Not necessarily, Ma'am. With a sand bottom the Danforth actually has a higher holding rating. If we really needed maximum holding power, we would use both anchors."

"Then why is it that we always make up the centerline anchor ready for letting go when we are at sea detail?"

Figure 6–1. Ground tackle on the forecastle of USS *Austin* (LPD 4). *U.S. Navy photo*

"Unfortunately, the port anchor has a nasty habit of hanging up, even after we have walked it part way out. I'd hate to have to depend on it if we needed to drop in a hurry, or for that matter when we're doing a precision anchoring like you will do day after tomorrow."

"Does the fact that we have two anchors but only one anchor windlass create a problem? For example, how do we do a Med moor?"[1]

"Well, my life would be easier if we had two windlasses, but we manage to make do. For a Med moor we would leave the centerline anchor on the windlass, and drop the port anchor on a compressor. It does make things a little slower in getting under way if we are using both anchors."

"Okay, I know the last shot of chain for each anchor is painted red, and the next to last shot is yellow, to give us warning when we are about to run out of chain. I also know that every fifteen fathoms there's a color code to let us know how much chain is out, but I can never remember the code without checking my wheel book."

"That's okay, Ma'am. We will let you know how much chain is out with regular reports. I have one request, though. If you have the conn when we are getting under way, we'd appreciate a little extra time when the chain is on the way in to repaint the chain markings. The paint tends to get chipped off in the chain locker."

"I understand. Any other requests?"

"Yes, Ma'am. In deciding how much chain to let out, we'd like to set our stoppers with a detachable link on deck. That way if we have to get under way in a hurry we can slip the anchor. We always use a watching buoy on the anchor, so that it will be easy to find if we have to leave it behind."

"Okay, Chief, I think that covers what I need. Thanks. I have one request for you. I always worry about the possibility of damage to our sonar dome. Please make sure to send me up continuous reports on the way the chain is tending, so that I can keep it away from the dome."

"Yes, Ma'am. I always do. And thanks for taking the time to talk the drill over in advance. It always goes smoother that way."

Capt. Don Rodriguez looked down at the meeting on the forecastle and smiled.

Two days later, *John Paul Jones* headed into San Diego Bay. Lieutenant (jg) James was on the bridge but had not yet taken the conn. Captain Rodriguez called her over to his chair and said, "Kathy, since this is the first precision anchorage any of us has done for a while, I'd like you to go over with me one more time how you plan to go about it."

"Yes, Sir. We will be anchoring with the centerline anchor, which is made up for letting go. There isn't much wind today, so I plan to make my approach to our assigned anchorage directly into the current. My computation of the tide is that it will be close to maximum ebb, which should be around a half knot in the vicinity of our anchorage, and the navigator concurs. I'd like to have as long a straight in approach to the anchorage as possible, but that is constrained both by the harbor configuration and other ships at anchor."

Captain Rodriguez interrupted. "Are you planning to drop when the navigational picture shows we are at the center of the anchorage?"

Lieutenant (jg) Jones smiled. "No, Sir. As best as I have been able to calculate, here on the bridge we are about sixty yards aft of the bullnose. All of our sensors are centered on the bridge, so I need to make an allowance."

"Good thinking," said the skipper. "For most purposes that might not make much difference, but we might as well give ourselves every edge we can. Tell me how you plan to get lined up for your approach."

"Yes, Sir. I want to head into the tidal current, which I estimate to be from about 350. I hope to confirm that once we are able to see if there are other anchored ships in the basin and how they are oriented. Then I'll head for a spot as far down current as room permits. I'd like to have at least a thousand yards of straight in shot for my approach. The tricky part is going to be judging my turning point so that I wind up heading toward the center of our anchorage and directly up current."

"How are you going to decide when to drop anchor?" asked Captain Rodriguez.

"I'd like to make the approach fairly hot, holding five knots until we are within about two hundred yards of the anchorage. I figure that if I back down one-third on both engines at that point that the bow should just pass through the center of the anchorage before gathering sternway. I'd like to drop just as the ship begins to move astern."

"Okay," said the skipper. "How are you planning to determine that point?"

"Well, Captain, you always tell us to use multiple sources of information. I figure that when the quickwater starts moving toward amidships we are dead in the water. I have a stack of wood chips I plan to throw over the side, and I'm going to have the lee helm singing out speeds from the pit log as we slow."

"Those are all good, Kathy" responded the captain, "but remember that all three show you your speed through the water, and what you want here is to know your speed over the bottom. The current is going to affect that."

"Yes, Sir, I hadn't thought of that. The GPS will give me an indication, but I'm not sure how much confidence I have in any single source for accurate speed readings at very low speeds. I guess I'd better plan on dropping just as the other indicators show that we are dead in the water, on the assumption that the half knot of ebb current we have calculated will provide just about the right amount of sternway to lay the chain out nicely."

"Sounds good to me, Kathy. As soon as we pass Ballast Point, and you are ready, go ahead and take the conn from Lieutenant Cover."

Lieutenant (jg) James took the conn as *John Paul Jones* passed the sub base, and for once everything went according to plan. After stoppers were passed on the chain, the navigator announced that they were within fifty yards of the designated anchorage.

"Well done, Lieutenant," the captain said. "Looks like we are ready to go for score next week."

The term "ground tackle" (pronounced *tay*-cull) refers to all of the equipment used in anchoring. It includes the anchor itself, the anchor cable (by convention the anchor chain is referred to as "cable"), connecting fittings, the anchor windlass, and all of the various hardware used to anchor, moor, or secure the anchor. The anchor and all of its associated gear is an essential part of the shiphandler's kit of tools.

Anchoring

Anchors are designed to dig in with a horizontal pull and to break themselves out with a vertical pull. Thus when anchoring the ship, we want to have some way on (sternway, if possible) to lay out the chain to put a horizontal pull on the anchor. To properly set the anchor, this pull needs to be as smooth, steady, and unidirectional as possible. No appreciable strain should be put on the chain until enough has been paid out to make the pull on the anchor horizontal. The normal technique is to pass a short distance over the intended

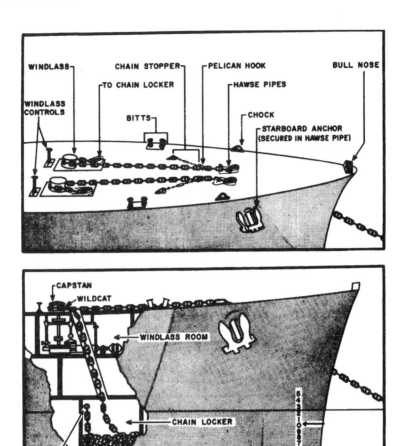

Figure 6–2. A typical ground-tackle arrangement. *Naval Institute Press*, Bluejacket's Manual, *22nd ed. Used with permission.*

point of anchoring, heading into the wind and current, then gather slight sternway to back through the same point, dropping the anchor as the bow again passes over the intended anchorage. Naval vessels with bow mounted sonars must take care that the chain is not allowed to touch the sonar dome.

Navigational circumstance permitting, it is desirable to approach the anchorage headed directly into the prevailing wind or current, whichever is strongest. If other ships are anchored in the vicinity, it is easy to observe how they are lying to their anchors, and make your approach in parallel. This

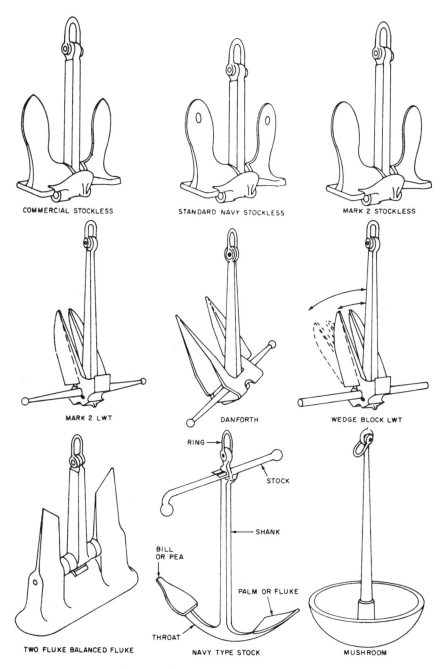

Figure 6–3. Types of anchors. *Naval Institute Press*, Bluejacket's Manual, *22nd ed. Used with permission.*

permits more precise control during the approach and facilitates setting the anchor properly. If you are to remain anchored for more than a brief period, after the desired amount of chain is out, the anchor can be set more firmly by backing down gently against the chain with the ship's engines. This needs to be done with some caution, since too much power can serve to break the anchor out.

The amount of chain to pay out when anchoring depends on depth of water, the length of time the ship will remain in that anchorage, the anticipated weather, and the character of the bottom. The most often cited rule for scope of chain when anchoring is five to seven times the depth of water. All other things being equal, the greater the scope of chain, the greater the holding power. In water deeper than fifteen fathoms most ships do not have enough chain to provide a scope of seven or more times the depth. In deeper water, it is usually preferable to set a second anchor than to transfer chain to achieve a longer scope. When anchoring in deep water, before letting go the chain should be walked out on the windlass until the anchor is close to the bottom. Otherwise the weight of the chain and anchor may overstress the brake and could even lead to the loss of the anchor and cable. It is always wise to position the chain so that a detachable link is in an easily accessible position on deck, in case it becomes necessary to slip the anchor on short notice. Anchors should always be buoyed to facilitate their recovery in case they must be slipped.

Anywhere other than the most protected harbor, the bow of an anchored ship will move around a great deal. This tendency can be damped by using a greater scope of chain or by putting a second anchor underfoot. The second anchor is dropped with a minimum scope of chain and serves to dampen the back and forth motion of the bow. If further stabilization of the bow is needed, this second anchor can be picked up and redropped when the ship is at maximum swing. The scope should then be adjusted to equalize the tension on both anchors. Even if the second anchor drags because of its relatively short scope of chain, it will serve to reduce the motion of the ship, and thus the shock load on the first anchor.

A ship anchored in a current can sheer from one side to the other, being brought up sharply as she reaches the limit on each side of the sheer. This jerking motion can cause the anchor to drag or to break out. The ship's rudder may be used to move the ship to one side of the sheer, and hold her there. The use of the rudder while anchored in a current may also be used to try to move your ship out of the way of another ship dragging anchor toward you. In any circumstance of strong current, or weather, of course, it is prudent to maintain a capability for getting under way on short notice.

In severe weather the strain on the ground tackle can be eased by steaming in place. The aim is to reduce the strain, not eliminate it. Slacking of the chain, followed by bringing it up taut, places a shock load on the anchor and makes it more likely that the anchor will drag or break out entirely. If the weather is severe enough to consider steaming in place, then a full watch should be set, capable of getting the ship under way should the anchor drag.

Anchor as a Shiphandling Aid

Whenever under way in restricted waters, the anchor should be kept ready for letting go. Not only can the anchor serve as a valuable aid in maneuvering, it also has saved many a mariner from embarrassment. The notion of the anchor as the ship's "emergency brake" can be misleading, however. Massive as the ground tackle may seem, it is not up to the task of bringing the thousands of tons of a ship making good way to a sudden stop. If it becomes necessary to use the anchor to slow or stop a ship making five knots or more, there are two alternatives. One is to drop the anchor only to short stay, so that the anchor drags over the bottom but does not dig in. The other is to continue to pay out chain until the ship is sufficiently slowed to make it safe to snub her.

On ships without bow thrusters or auxiliary propulsion units the anchor is the only way the shiphandler can exert force directly on the bow without using a tug. For example, naval vessels typically have more sail area forward than aft, making it difficult to twist the bow up into the wind. It is not unusual on a windy day to find yourself being blown sideways across a harbor while unable to bring the bow into the wind. In these circumstances dropping the anchor at short stay will bring the bow up smartly into the wind, salvaging in seamanlike fashion what could have turned into an uncomfortable situation.

The anchor can also be used to control the bow when being set onto a pier by wind or current and has often been called the "poor man's tugboat." To do this, the anchor on the side away from the berth should be made ready for letting go. The technique is to make your approach somewhat wider than normal, dropping the anchor about one hundred yards from the intended final position of the bow, veering chain as necessary to keep the bow under control. The retarding effect of the chain permits the use of more power ahead on the engines, increasing the effect of the rudders and therefore your control of the stern. Thus the ship can be eased on to the pier even with strong onsetting forces from wind and current. Once tied up the chain is slacked to lie on the bottom.

When ready to get under way, heaving in on the chain will walk the bow away from the pier. Engines can be used to twist the stern away from the pier, or worked ahead against the retarding effect of the anchor, giving more effective rudder control. A potential difficulty arises if the anchor is foul and cannot be cleared expeditiously. This leaves the ship in an awkward position. It is therefore even more important than usual that the anchor be buoyed and a detachable link positioned for slipping the anchor if necessary. Since the water in the vicinity of the pier will usually be relatively shallow, recovery of a properly buoyed anchor is not difficult.

The anchor can also be of great assistance when backing into a berth under conditions where wind or current present a problem. The ship is lined up to back into the assigned berth, and starts to back. About two hundred feet before arriving in final position, with sternway on, the anchor away from the pier is dropped. Chain is veered as needed, keeping the bow under control. Ship's engines and rudder are used to control the stern. Once tied up alongside, the chain is slacked to the bottom.

The Med Moor

The Mediterranean (Med) moor is a method of mooring a ship using two anchors to secure the bow, with a stern line or lines to the pier. As the name suggests, it is most often used in the Mediterranean, primarily to make best use of limited pier space in crowded harbors. It is a strong moor, and it has an advantage over nesting in that each ship has its own brow to the pier. Its primary disadvantage is the possibility of fouling anchors with adjacent ships. The danger of fouling can be reduced somewhat if circumstances permit ships to get under way in reverse order from the order in which they moored.

As with all shiphandling, the key to a Med moor is planning of the maneuver. The goal is to place the anchors about 100 to 150 yards apart on a line parallel to the pier and equidistant on either side of the position the ship will occupy when moored (see fig. 6–4). The distance from the pier the anchors should be placed is the length of the ship, plus a distance comfortably less than the shortest chain to be used. If the shortest chain is 105 fathoms, plan for using perhaps 75 fathoms (150 yards). To allow for the fact that each chain will be at a 20- to 30-degree angle, allow for 120 yards off of the pier in addition to the ship's length. For example, if the ship is 540 feet (180 yards) in overall length, the anchors should be dropped on a line approximately (180 plus 120) 300 yards from the face of the pier. These

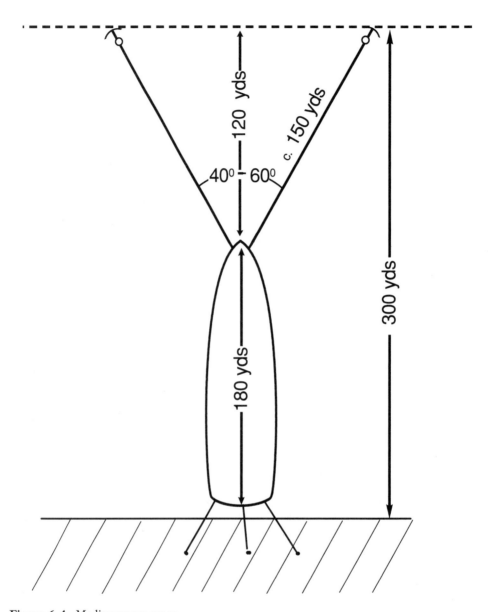

Figure 6-4. Mediterranean moor.

planned positions should be marked on the harbor charts so the navigation team can aid in the approach.

The approach is made approximately parallel to the pier and toward the point chosen for the first anchor, which should be the one on the offside, away from the pier. Reduce speed to bare steerageway as the drop point is approached, dropping the anchor at the chosen point, about fifty yards before the bow comes abreast of the assigned point on the pier. (If the ship has a single wildcat, it should be used for this first drop, with the compressor used for the second anchor.) As the anchor is dropped, full rudder is used to head toward the second drop point and to keep the chain clear of the ship, augmented by a twist if necessary. Chain to the first anchor is veered as the ship heads toward the second chosen drop point. After passing slightly through the point, the second anchor is dropped with slight sternway. As the stern is twisted toward the pier, chain is veered or taken in as necessary to locate the bow at a point 90 degrees from the point on the pier at which the stern is to be located. This need not be too precise at this point, as it can be further adjusted after the ship is moored. As the ship comes into line, it is backed gently against the catenary of the chain to locate her at the desired distance from the pier. It may be desirable to move the conn to the fantail for this part of the evolution. Once the stern line is secure the anchors are adjusted to obtain enough tension to keep the stern safely off the pier. It is important that the stern line not be too vertical to avoid overstress from water surge or tidal variations. If a line through the stern chock would be too vertical, lines may be run out of quarter chocks to get a longer lead.

In getting under way from a Med moor, the last anchor placed should be retrieved first. It is a good idea to retain a stern line as long as possible to keep the position of the stern under control as the anchors are retrieved. Special care should be taken to avoid any possibility of getting the line into your screws. Because of the possibility of a foul anchor, it is prudent to have a tug standing by to help until the "anchor clear" report is received.

A variation on the Med moor is to moor with a stern wire to a buoy and two anchors placed as discussed above. This is a strong moor and has the advantage in a crowded harbor of not requiring the additional room to swing that simply anchoring requires. Another variant is the use of a single anchor, with the stern to the pier. With single-anchor ships like the *Perry*-class frigates this may be a necessity. It is not, however, a strong moor, and in some ports "loaner" anchors and chain are available to strengthen the moor. Yet another alternative is to moor two single-anchor ships alongside each other, with one ship's anchor to port, the other's to starboard, and both sterns moored to the pier.

Mooring to a Buoy

Mooring buoys are not as prevalent as they once were but still may be encountered in some ports frequented by naval vessels. Mooring to a buoy has two advantages over anchoring. One is that mooring buoys almost always are more securely anchored than can be achieved with the ship's own anchors. The other is that a properly anchored buoy remains nearly stationary, reducing the swinging arc of the ship. The downside of mooring to a buoy is that it is generally more difficult than anchoring, requires the use of a boat, and can involve greater hazard to the personnel engaged in the evolution.

To moor to a buoy an anchor is detached from its chain, and the chain is led out through the bullnose of the ship and shackled to the buoy. Because the weight of the chain precludes manhandling it into position, the preferred technique is to first attach a wire or springlaid buoy line to the buoy and use this temporary attachment as a trolley from which the chain is supported by shackles (see fig. 6–5). Once the trolley wire is in place, the chain is slid down the trolley to the personnel on the buoy to connect the chain to the buoy with a mooring shackle.

As is evident, this evolution requires precise shiphandling. If at all possible, the approach to the buoy should be directly into the prevailing current or

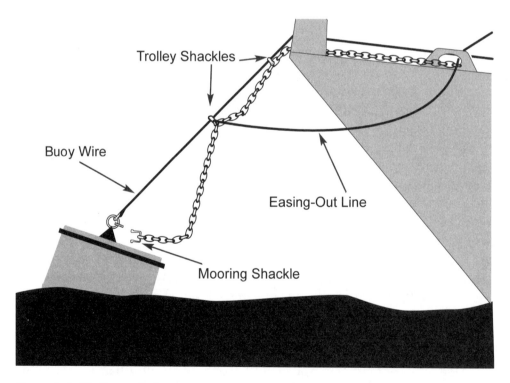

Figure 6–5. Trolley method.

wind. An approach from cross current makes it almost impossible to hold the ship's bow in the needed position long enough to complete the moor, unless bow thrusters, auxiliary propulsion units, or a tug can be used.

Even more than with most evolutions, careful preplanning and preparation are necessary. The pre-planning meeting should include a briefing of the buoy party, the boat crew, and the forecastle crew. The boat, with the buoy party, should be in the water before starting the approach to the buoy. At two hundred to three hundred yards, the boat should come alongside the bow to pick up the messenger line for the trolley wire, then proceed ahead of the ship to place the buoy party with the messenger line on the buoy. All members of the buoy party should wear life jackets.

Depending on the weight of the trolley wire, the messenger can be led through the buoy ring and back to the ship to use the windlass to haul the trolley wire to the buoy. The buoy party must be removed from the buoy before taking any strain, and then may be placed back on the buoy when needed to

Figure 6–6. USS *Puget Sound* (AD38) moored to a buoy in Gaeta, Italy, with USS *John Hancock* (DD 981) tied up alongside. *U.S. Naval Institute photo archives*

complete the hookup. With the ship's bow positioned over the buoy, the chain is then slid down the trolley wire and shackled to the buoy.

The approach to the buoy needs to be well controlled because of the potential hazard to the personnel on the buoy. Any abrupt strain or even a light contact of the buoy with the bow could dump the buoy party into the water. The boat crew must be instructed not to allow the boat to get between the buoy and the ship. For ships such as aircraft carriers or large deck amphibious ships in which the buoy is not visible from the bridge when the ship is in position, it is a good idea to shift the conn to the forecastle to achieve the needed delicacy of control. For destroyers or cruisers it is desirable to place the buoy a few yards off the bow to permit the conning officer keeping it in sight from the bridge. In the approach to the buoy, it should be kept at a constant bearing, and any sternwalk when backing should be anticipated. Use of engines during the hookup process should be kept to a minimum. Once moored, the load placed upon the buoy can be reduced by veering chain, so that the catenary of the chain dampens any tendency to jerk on the buoy.

In some circumstances it may be required to moor to buoys fore and aft, using chain forward and wire aft. This is just a matter of taking one buoy at a time, starting with the up-current buoy. Once that is connected, the chain or wire is slacked off to permit moving the ship into position to moor to the down-current buoy.

As with line handling, actions with the anchors take place at a location remote from the conning officer on the bridge. It is therefore again of particular importance that both commands and reports be clear and uniformly understandable:

Command	Meaning
"Make the (port) anchor ready for letting go."	Designates the anchor to be prepared for letting go.
"Dip the (port anchor)."	Suspend the anchor by wire under the bullnose (less likely to damage ship alongside).
"Walk the anchor (to the water's edge, fifteen fathoms on deck, etc.)."	Use the anchor windlass to walk out the anchor under control to the point indicated.
"Veer to (forty-five fathoms on deck)."	Pay out chain to the ordered amount.

"Let go."	Trip the pelican hook, releasing the stopper so the chain can run freely.
"Heave around to short stay."	Walk the chain in on the windlass to the point that the anchor is just about ready to break ground.
"Avast heaving."	Stop heaving in on the anchor chair.

When working with the anchor and chain, the forecastle makes frequent reports on the amount of chain out, the strain on the chain, and the angle relative to the bow, given in terms of clock direction. A typical report would be "Forty-five fathoms on deck, chain tending at five o'clock; light strain." Strain is reported as light, moderate, heavy, or no strain.

Report	Meaning
"Chain tends (three o'clock)."	The anchor chain tends from the hawse pipe on the indicated relative clock bearing from the bow.
"(Fifteen) fathoms at the water's edge."	The amount of chain out at the point indicated, usually "on deck," "at the hawse,"or "at the water's edge."
"Anchor is at short stay."	Anchor is nearly up and down, and just before beginning to break out.
"Anchor is up and down."	Anchor has broken ground, but is still resting on the bottom.
"Anchor breaking ground."	Anchor is about ready to break out. Can be observed because the chain is nearly vertical and is jerked as the anchor breaks ground.
"Anchor's aweigh."	The anchor is clear of the bottom, and the ship is now under way.
"Anchor in sight."	Anchor detail can see the anchor.
"Anchor is clear (or foul)."	Clear: there is little bottom debris clinging to the anchor. Foul: the anchor is hooked on to some bottom obstruction, such as a cable or another ship's chain.
"Anchor is secured for sea."	Anchor is housed in the hawse pipe, and the stoppers have been passed.

"Anchor is ready for letting go."   The anchor windlass brake is con-
                                                 nected, all but one of the chain stoppers
                                                 have been removed, and the weight of
                                                 the anchor is hanging on the remaining
                                                 stopper so that releasing the pelican
                                                 hook will cause the anchor to fall.

A standard "shot" of anchor chain is fifteen fathoms (ninety feet) in length. Shots are connected by detachable links. The detachable links are painted red, white, or blue to show how much chain is out. The number of adjacent links painted white identifies the shot number. All of the next to last shot is painted yellow, and all of the last shot of chain is painted red, to provide a visual indication that almost all of the chain is out. The anchor is normally attached to the chain with a short "bending shot" that incorporates a swivel. The color coding of the chain is as follows:

Shot Number	Color of Detachable Link	Number of Adjacent Links Painted White
15 fathoms	Red	1
30 fathoms	White	2
45 fathoms	Blue	3
60 fathoms	Red	4
75 fathoms	White	5
90 fathoms	Blue	6
105 fathoms	Red	7

Using Ground Tackle

Naval vessels, perhaps because of their generally greater power and maneuverability, tend to make less use of their anchors than do merchant ships. Properly used, a ship's ground tackle can be the shiphandler's best friend, permitting evolutions not otherwise possible without the use of tugs, and providing a means of salvaging the situation when something goes wrong. No shiphandler's education is complete until he has developed skill in the use of the anchor and all of its associated gear.

7

TRANSITING THE CHANNEL

---◄o►---

Captain Santamaria turned to the pilot and said, "Very nice, Captain O'Neill, and now I'd like to give Lieutenant Wright an opportunity to conn the ship for the channel transit."

They had just walked into the pilot house of USS *Bataan* (LHD 5) from the port bridge wing, from where the pilot had conned the big amphibious assault ship away from Pier 4, Norfolk Naval Base. The two tugs had just been released. Now the 40,000-ton behemoth lay dead in the water midchannel in the Norfolk Harbor Reach, waiting patiently for direction.

Captain O'Neill turned to the young officer who had been standing with him and following close as the pilot had moved between bridge wings. "Okay, Lieutenant," he said. "Ready to take the conn?"

"Yes, Sir." Tom turned to the center of the pilot house. "This is Lieutenant Wright, I have the conn." A half dozen sailors acknowledged the announcement as the new conning officer turned and saluted the commanding officer. "Captain Santamaria, I have the conn, Sir."

Returning the salute, Santamaria said, "Very well, Tom. Take us out to sea."

Tom looked at the pelorus and ordered, "All engines ahead two-thirds, indicate turns for ten knots. Left 10 degrees rudder. Come left to new course zero zero four." The helmsman and lee helmsman responded with exactly Tom's orders just as the navigator sang out, "Channel heading 004. Speed limit ten knots. Hold you 20 yards right of midchannel; 1,150 yards to next turn." Tom acknowledged with

"Conn, aye" to the navigator and "Very well" to both helm and lee helm as *Bataan* started its channel transit to sea.

A minute later the phone talker chanted, "Combat holds you twenty-five yards right of track, recommends course 002," and Tom acknowledged. He turned and looked over the shoulder of Lt. (jg) Gary Erickson sitting at the Electronic Chart Display and Information System. "What say ECDIS?" Tom asked. Erickson replied, "ECDIS concurs." Then Tom heard "Navigator concurs. Recommend steer 002. Five hundred yards to 'rudder over.'"

Tom ordered the minor course change as the phone talker called, "Port Lookout reports merchant ship off port beam about five miles, closing."

Tom went out to the port bridge wing and put to his eyes the binoculars that had been hanging around his neck. "Container ship in Newport News Channel," he said. He turned to the operations specialist at the Automatic Radar Plotting Aids radar. "ARPA, can you give me a course and speed on that contact?" "Yes, Sir," the OS replied, manipulating his controls. He then sang out, "Contact Delta bearing 295, range 3,700 yards, on course 049, speed fourteen knots. CPA 350, 950 yards. Time of CPA 3.5 minutes."

"Captain, I'm going to make sure he goes on ahead of us." Jim pointed to the black container ship on the port beam. As the captain nodded, Jim ordered all engines ahead two-thirds with turns for eight knots and turned to Lt. (jg) Phil Bozzelli standing in the pilot house holding the bridge-to-bridge radio handset. "Got him, Phil? Black hull container ship, course, speed? Let him go on ahead."

"Got it all," Phil replied as he checked his notes then spoke into the radio handset. "Calling black hull container ship outbound in Newport News Channel on course 049, fourteen knots. This is Navy warship number 5 on your starboard bow outbound in Norfolk Harbor Reach, on course 002, bearing 115, one and a half miles from you. I am slowing to eight knots. Intend to let you go ahead of me. Navy warship 5 on channel 16. Over."

"Navy warship 5 this is *Casaba Moranga*. Concur. Thank you. Out."

Tom nodded at Phil and turned to the navigator. "Distance to rudder over? ECDIS?"

"Fifty seconds to rudder over at Buoy 5. New course 026," the navigator replied. ECDIS concurred, and Tom studied the channel ahead. As *Bataan* passed Buoy 7 to starboard, approaching Buoy 5, he took a look at the merchant ship now on his port bow and moving ahead and went out on the port bridge wing. From the phone talker Jim learned that combat concurred with both navigator and ECDIS.

"Right full rudder. Steady on course 026." His helmsman responded and Tom acknowledged as the big ship came slowly to the right, around Buoy 5, and steadied on the new heading. *Casaba Moranga* had proceeded ahead and was in the Norfolk Harbor Entrance Reach. Tom pointed at the stern of the container ship. "ARPA, looks like he increased speed."

"That's affirmative. Contact Delta is on course 045, making eighteen knots."

"Navigator, can I increase to twelve knots now?"

"That's affirmative. Twenty seconds to rudder over. New course will be 045 with Buoy 3 abeam to starboard."

Tom had gone back into the pilot house. As he approached Buoy 3, he ordered, "All engines ahead two-thirds. Indicate turns for twelve knots," then "Right full rudder. Steady on course 045."

Bataan completed the short turn into Norfolk Harbor Entrance Reach, passed Buoy 3, and the navigator reported, "Hold you on track, midchannel in outbound channel. This is a long leg, Tom, 3,960 yards to next turn just past Fort Wool to starboard and Old Point Comfort to port, and before Thimble Shoals. You'll be set by wind and current to starboard as we clear Old Point Comfort."

Tom acknowledged and remembered that same information from the navigation brief two days ago as he checked his notes and the small chartlet he had made. He altered course 2 degrees to port to compensate for wind and current. Everything was going according to plan.

ARPA had been tracking a contact in the channel far ahead when Tom Wright and Captain Santamaria picked it up with binoculars. "Looks like he just turned in to Thimble Shoals Channel," the captain said. "What say you, ARPA? ECDIS?"

Both reported Contact Echo had just changed course and they had new course, speed, and CPA, and combat concurred with the information. Tom knew that *Bataan* would be changing course soon and the CPA would change.

"Looks like a red-hulled tanker in Thimble Shoals Channel, Phil. Can you tell him I intend a port-to-port passage?"

"Sure can, Tom. Let me get some data." Phil queried the ARPA and ECDIS operators.

"Calling red-hull tanker inbound Thimble Shoals Channel on course 288, speed sixteen knots, vicinity Buoy 13. This is Navy warship 5 on your port bow outbound in Norfolk Harbor Entrance Reach, on course 045, speed twelve knots, bearing 302, four and a half miles from you. I intend port-to-port passage. Navy warship 5 on channel 16. Over."

A British accent came back with "Navy warship number 5 this is *Birmingham Lady.* I see the same, port-to-port. Thank you. Out."

With Old Point Comfort abeam to port, Tom put the rudder over and *Bataan* turned slowly to 076. They waved to the bridge watch of inbound *Birmingham Lady,* and a short time later, with recommendation of the navigator and concurrence of ECDIS, Tom changed course to 092 and then 108 as he entered Thimble Shoal Channel. He was in the final legs of the channel transit from Chesapeake Bay to sea.

As our ship approaches the entrance channel, the environment becomes increasingly complex. Traffic is heavier, shoal water is closer, navigational aids multiply, and currents become more significant. In preparation for this the ship has had a navigation brief or voyage conference to plan all of the details of the transit. The special sea and anchor detail is set, to put our best qualified people in all key assignments. The navigator and his team start plotting the ship's position at closer intervals. The conning officer slows her from speeds appropriate to blue water to the more deliberate speeds appropriate to the harbor.

Despite the presence of the navigation team, the conning officer bears a large part of the responsibility of keeping the ship in safe water. Before assuming the watch, he should have carefully reviewed with the navigator both the harbor chart and the *Coast Pilot.* The *Coast Pilot* contains useful information about the harbor that cannot be found on the chart. It is important that CIC, the navigator, and the conning officer each have their own harbor chart, with identical intended tracks, turning bearings, and danger bearings laid out. The chart should be carefully studied in advance, looking for shoals near the intended track, the locations of significant landmarks, ranges to be followed, any particularly tricky turns to be negotiated, narrow places in the transit, usable emergency anchorages, and so on.

At night the harbor becomes a more difficult place. City lights can make it harder to spot and identify navigational marks. Distances become more difficult to judge. Visual cues of all kinds are more difficult to come by. Dark adaptation is vital, and tight control is needed over any potential light sources on the bridge that could affect it. All stations need to be trained not to direct lights toward the bridge. It can be useful to man signal searchlights on both sides to light buoys and other navigational marks.

Piloting through the Channel

On a warship at special sea and anchor detail there are normally three teams keeping track of the ship's progress through the harbor. One is the navigator and his team, piloting primarily by visual bearings. Another is the team in combat information center, using primarily radar information. The controlling team is composed of the officer of the deck or other designated conning officer, the commanding officer, and a pilot, if on board. Only one person can have the conn, but the others serve as observers and advisors. This controlling team keeps track of the ship's position primarily by visual observation of ranges, buoys, and significant landmarks, supplemented by information from

Figure 7–1. Navigational range as represented in a shiphandling simulator. Own ship is off the range to the right. *Maritime Safety International*

ECDIS. All three teams make use of information from the fathometer and the global positioning system.

Most harbors have ranges to mark the center of the channel for critical parts of the transit. The ranges consist of a lower mark in front, and a higher mark placed some distance behind. When in line they serve as a very precise indicator of the center of the channel. If you are on a range it provides a good visual fix each time you pass a landmark or buoy abeam. Ranges vary in sophistication from simple unlighted structures to very precise laser ranges, such as that used for the narrow channel leading into Cairns, Australia. In most harbors used by naval vessels, the ranges will be lighted at night. A range provides a good opportunity to observe the effect of wind or current in the channel. If it is necessary to steer a course to one side or the other of the range course in order to remain on the range ("crabbing"), this is a good indication of any current that is acting athwart the channel. This does not, however, tell us anything about the effect of any current that may be running parallel to the ship's course. For this we must rely upon the navigator's calculation of set and drift, calculated in this instance by comparing our navigational speed over the ground with the speed rung up on the engines. The conning officer also needs to learn to estimate current by eye, particularly in the vicinity of turning buoys or the pier. Current creates a wake from buoys and pilings, as though they were moving through the water at the speed of the current. Buoys also often assume a list as the current pushes against them. Some caution is necessary here, since buoys also assume a list for reasons of their own, such as having taken on some water. Currents of one knot or less can be seen on a buoy, and as the current increases it becomes more evident, with a three-knot current showing a visible wake for several yards. In some channels it is not unusual to encounter this much current or more at maximum ebb or flood tide. For shiphandling purposes, three knots of tide can be considered as roughly equivalent to ninety knots of wind, depending upon a ship's sail area.

Narrow channels are normally marked by buoys on either side of the channel. This gives the conning officer an excellent reference to his position within the channel. Buoys are also frequently used to mark shoals and turns in the channel. Useful as the buoys are, they should not be relied upon as being precisely as charted, particularly after a storm. Because one buoy looks much like another, always use binoculars to read the number and positively identify each one. A hazard in the channel is that of being set down upon a buoy. This can be avoided by visually observing each buoy as it is approached.

To pass a buoy safely to starboard, the buoy needs to appear to be moving right against its background. If the buoy is not moving relative to the background, you are on a constant bearing and in danger of colliding with the buoy. This observation needs to be made from the engaged bridge wing, since in close quarters the buoy can have a drift as observed from the ship's centerline but still be on a steady bearing with the side of the ship. It is a bad idea to touch a buoy. For one thing, it probably means that you are getting too close to the edge of the channel. For another, if lucky, you may get only scratched paint, but the more serious hazard is the possibility of entangling the buoy's mooring gear in your screw. If you find yourself being set onto a buoy it is often possible to do a step-aside, swinging the ship's bow away from the buoy, followed by a rudder shift to swing the stern clear. Obviously this can be done only if the water depth in the vicinity of the buoy is adequate, since the maneuver moves the ship somewhat outside the buoyed channel. It is a better idea not to get that close to begin with.

Shallow-Water Effects

As the depth of water beneath the keel decreases, the ship begins to be affected. For large merchant vessels, MacElrevey reports that the turning radius in shallow water for a given rudder angle can be almost twice that for the same ship in deep water.[1] The degree to which this affects the ship depends upon water depth, speed, and the configuration of the individual ship, so no universal rule can be stated. The prudent shiphandler, however, will anticipate this and avoid the need for full rudder turns in shallow water. Shallow water can increase the sternwalk in a single-screw vessel. Stopping distance with a backing bell can also increase in shallow water.

A potentially hazardous shallow-water effect is the increase in stern draft referred to as "squat." Squat is a complex phenomenon caused by two things. One is the Venturi effect, in which the increased velocity of the water flowing around the ship's underbody creates a region of lower pressure under the keel, increasing the ship's draft. The other is the ship riding up on its own bow wave, raising the bow and lowering the stern. Squat occurs with high speeds in deep water as well as shallow, but the effect is both more pronounced and more of a problem in shallow water. As the speed of the ship increases, it settles lower in the water. This takes the form of an increased bow wave and stern wave, with depressed water alongside. As speed continues to increase, the ship begins to ride up on its bow wave and the stern sinks lower. A substantial "rooster tail" forms behind the ship. The amount of squat varies in proportion to the square of the speed. The *Arleigh Burke* class of guided-missile destroyers is reported to squat five feet at fifteen knots in fifty feet of water, seven feet at twenty knots, and as much as twelve feet at twenty-seven knots made good.[2] In water shallower than fifty feet, the effect can be expected to increase. Thus squat can increase the effective navigational draft of a ship, and can turn a perfectly adequate channel into one that is too shallow. In addition, the wake effects that accompany squat generate waves that can be damaging to moored ships and craft inside a harbor. The moral here is clear: slow down in shallow water.

Traffic

Traffic density becomes markedly greater as the ship enters the channel and the harbor. More traffic is likely to be encountered in the channel and inside the harbor in an hour than in days at sea, and each ship or boat is a potential hazard. The conning officer and his watch team, aided by CIC, need to track and evaluate all contacts. While it is obvious that the conning officer must attend to all contacts that could present a hazard, some prioritization may be needed. This is mostly a matter of common sense: close ships are of greater concern than those further away; a rapid closing rate more dangerous than slow, large ships more of a problem than small craft. As always, those with constant bearing and decreasing range require special attention. It is dangerous to fixate on a single contact while a hazardous situation develops with one or more others. Most well-ordered bridges have found it prudent when at special sea detail to assign a qualified officer to remain on the bridge wing away from the rest of the conning party to avoid an unobserved problem developing on the unengaged side.

The key to safe passage with harbor traffic is communication, and there are several ways to communicate. One of the most important is simply to maneuver so as to make your intentions visibly clear to another vessel. For example in a meeting situation a course change of only a degree or two might be sufficient for passage at a safe distance. It is, however, much better to "signal with your bow" and make a sufficiently large course change to be unambiguously observable, then return to base course after the situation has clarified.

For whatever reason, naval officers seem to be reluctant to use the ship's whistle. There is absolutely no reason not to parallel VHF communications with the use of the ship's whistle. In addition, the Inland Rules have provisions for paralleling whistle signals by light.[3] There also seems to be a prevailing belief that the use of the five short blasts signal is restricted to situations of immediate danger. In fact, the signal is entirely appropriate whenever you are in doubt about the action of an approaching vessel, when you think it is doing the wrong thing, when meeting or overtaking signals are crossed or ignored, or any other circumstances in need of clarification. It is particularly useful to remind small craft of the provisions of Rule 9 (b) of both International and Inland Rules, that a "vessel of less than 20 meters in length or a sailing vessel shall not impede the passage of a vessel which can safely navigate only within a narrow channel or fairway."

Bridge-to-Bridge Radio Communications

VHF radio is now almost universally used for communication with other vessels. It can be of great value in clarifying intent, but it can also be a source of confusion. VHF radio can be heard over a considerable distance, and it is easy to think you are talking to one vessel when it is really another that may not even be in sight. When calling on VHF to exchange intentions, do not say things like "Ship on my port bow." Rather, call "SeaLand container ship outbound in Thimble Shoals channel this is U.S. Navy destroyer hull number 52 inbound on your port bow" or some other way of unambiguously identifying the two ships.

Once communications are established, it is vital that the information communicated be clearly understood. English is becoming the standard maritime language. In 1973, the Maritime Safety Committee of the International Maritime Organization agreed that where language difficulties arise, a common language should be used for navigational purposes, and that language should be English. Subsequently, the IMO developed "Standard Marine Communication Phrases," which were formally adopted in 2001 and published in 2002. Unfortunately, many of the things you may want to say in communicating with another ship in the channel are not covered by the standard phrases.

Perhaps more useful is an understanding of the principles that underlie the IMO SMCP. It builds upon a basic knowledge of English, reflecting widespread maritime English language usage in ship-to-shore and ship-to-ship communication. In the main, function words such as *the*, *a/an*, and *is/are* are omitted. Wherever possible, synonyms and contracted forms are avoided.

In response to yes/no queries, fully worded answers are used. The phonetic alphabet specified is identical to the NATO phonetic alphabet already used in the Navy. Numbers are pronounced as follows:

Number	Pronunciation
0	ZEERO
1	WUN
2	TOO
3	TREE
4	FOWER
5	FIFE
6	SIX
7	SEVEN
8	AIT
9	NINER
10	TOUSAND

When the answer to a question is in the affirmative, say "Yes" followed by the appropriate phrase in full. When the answer to a question is in the negative, say "No" followed by the appropriate phrase in full.

Even with a working knowledge of the Standard Maritime Communication Phrases, not all officers on ships you may encounter will have a perfect command of English. Most U.S. ships, particularly under the Inland Rules, tend to discuss meeting situations in terms of their intentions for meeting and passing. Foreign flag vessels, more accustomed to the International Rules of the Road, tend to discuss how they intend to change course. To avoid the possibility of confusion that would result if the other bridge picked up only "port" in your transmission, it is wise to state your intentions both ways: "I intend to alter course to starboard to pass you port to port." If you suspect any possibility of confusion, in inland waters you can add "for a one whistle passage."

In the United States, the Coast Guard is the authority for frequency allocation, proper procedures and use of radio equipment under the Vessel Bridge-to-Bridge Radiotelephone Act by all ships, including Navy ships, so use should be in accordance with Coast Guard regulations. Navy ships can talk to each other on Navy radio circuits using Navy terminology, but when using a Coast Guard circuit to talk with a merchant ship, the Navy ship should follow Coast Guard procedures.

Compliance with these procedures by Navy ships meets international standards of seamanship so it is important that officers using the VHF bridge-to-bridge radiotelephone on the bridge of a Navy ship, typically channel 16, use proper terminology when calling a merchant ship. Coast Guard regulations include recommended formats for ships initiating communications on a designated navigational frequency. These formats should follow a general voice radio procedure, but in practice these formats and procedures are not carried out exactly, and it should be understood that the Coast Guard offers them with the caveat that terminology should be "similar to" that provided by regulations. The recommended format for use of bridge-to-bridge radiotelephone is as follows (for vessel initiating call-up):

1. Identify other ship. Name if possible (otherwise, description such as hull color and type vessel), course, speed, vicinity of (and/or where bound). This should be done in such a manner that the mate on watch in the other ship easily understands "That's me being called."

2. Identify own ship. Navy hull number, course, speed, vicinity of (and/or where bound), and position of your ship with reference to other ship. It is most important that you give the position of your ship *from the other ship*, not vice versa. Do not say, "Ship on my port bow." You should identify your ship so that the mate on watch in the other ship can look in the direction you gave and say, "That's the Navy ship calling me." For security purposes, it is best not to use your Navy ship's name and hull number in the same transmission or sequence of transmissions.

3. State intention, proposal, or plan: meeting side-to-side, crossing with reference to other ship's bow or stern, or overtaking (passing) with reference to other ship's side. If you know what you want to do, use "I intend to," followed by your intention. If you want to get the other ship's concurrence, use "I propose" or "I plan." Do not ask, "What are your intentions?" State what you intend or give your proposal or plan. Remember, all merchant ships use miles, not yards, so when talking to a merchant ship use miles.

The vessel acknowledging receipt of the message answers with "(name of vessel calling. In this case, 'Navy warship [number]'). This is (name of vessel answering). Received your call," and follows with a concurrence and/or indication of intentions.

Some examples of VHF bridge-to-bridge radiotelephone formats follow:

1. Meeting in a channel: "Calling red ro-ro ship heading outbound San Diego at ten knots vicinity Buoy 22. This is Navy warship

number 65 off your port bow, inbound San Diego at ten knots located vicinity Buoy 12. I propose port-to-port meeting. Over."

2. Crossing in a channel: "Calling black hull container ship three miles off Ocean View proceeding north Chesapeake Bay Channel. This is Navy warship number 10 on your port bow, located Hampton Roads, three miles south of Old Point Comfort on course 075, speed ten knots. I intend to cross astern of you. Over."

3. Overtaking in a channel: "Calling *Wonton Maru* proceeding south Thimble Shoals Channel. This is Navy warship number 970 located two miles astern of you. I plan to overtake you on your starboard side. Over."

4. Sailboat in channel: "Calling white sailboat *Sea Tern* heading northeast three knots in midchannel San Diego, vicinity Buoy 16 near North Island. This is Navy warship number 4 located one mile astern of you, on course 075, speed seven knots. Request you move to side of channel so that I can remain in channel. Over."

5. Meeting at sea: "Calling blue hull fishing vessel five miles northwest of Race Point Lighthouse on course 300, speed eight knots. This is Navy warship number 55 located four miles off your starboard bow on course 120, speed fifteen knots. I propose starboard to starboard meeting. Over."

6. Crossing at sea: "Calling black hull tanker on course 120, speed twelve knots. This is Navy warship number 49 located four miles on your port bow on course 240, speed sixteen knots. I intend to cross astern of you. Over."

7. Overtaking at sea: "Calling white container ship six miles northeast of Cape Henry Light on course 280, speed ten knots heading toward Chesapeake Bay. This is Navy warship number 76 on your starboard quarter on course 280, speed eighteen knots heading toward Chesapeake Bay. I plan to overtake you on your starboard side. Over."

Traffic Control Systems

One of the few places in which there is a substantial difference between the International and Inland Rules of the Road has to do with what the International Rules call "traffic separation schemes" and the Inland Rules call "Vessel Traffic Services" (VTS). The International Rules' traffic separation schemes are passive systems intended to bring order to heavily navigated channels by establishing corridors in which all traffic is intended to move in the same direction. The Inland Rules' Vessel Traffic Services are managed systems with shore-based managers using an array of electronic aids.

International Rule 10 recognizes traffic separation schemes adopted by the International Maritime Organization, an agency of the United Nations. Proposed schemes within IMO-established guidelines are presented by the country or countries concerned in a process that leads to formal adoption. Once adopted, the separation lanes are shown on navigation charts. A typical scheme divides the channel leading to a harbor into an inbound lane and an outbound lane, often with buoys to mark the division between the two lanes.

Although the intent of International Rule 10 is to reduce the risk of collision in heavily trafficked channels, it states specifically that a traffic separation scheme "does not relieve any vessel of her obligation under any other rule."[4] The Rules offer three principles for operating in a traffic separation scheme. One is to "proceed in the appropriate traffic lane in the general direction of traffic flow for that lane."[5] This has been described as the "go with the flow" rule. The second rule is "as far as practicable keep clear of a traffic separation line or separation zone." This might be called the "don't hug the centerline rule."[6] The third rule says normally to join and leave the separation scheme at the end of the lane, but that if you have to join or leave at other points to do so "at as small an angle to the general direction of traffic flow as practicable."[7] Think of this as easing into or out of traffic, similar to merging onto an interstate highway from an on ramp.

An additional rule is provided for vessels having to cross the traffic separation channels. After urging avoidance of crossing, if possible, Rule 10 (c) says if you have to do it anyway, then do it at as close to a right angle as practicable. The intent here is to make it clear that the vessel is crossing, not part of the separation flow, and to minimize the amount of time a crossing vessel is in the designated channels. Note that nothing in the Rules changes the normal rules for crossing situations as to which ship is the stand-on vessel and which is the give-way vessel.

International Rule 10 also provides for the establishment of "inshore traffic zones." These are primarily intended for vessels less than twenty meters in length, sailboats, and fishing boats. Larger vessels are to stay out of these zones except in emergencies. The intent is to the extent possible to separate the larger faster ships from small craft, with the larger vessels using the established traffic lanes. The Rules also urge that to the extent possible ships should join and depart the established lanes only at their terminations and to use particular caution at these points.[8] Vessels not using the separation scheme should stay as far away from the established channels as practicable.[9]

In contrast to the passive traffic separation schemes under the International Rules of the Road, the Inland Rules provide for actively managed systems. Rule 10 in the Inland Rules reads in its entirety, "Each vessel required by regulation to participate in a vessel traffic service shall comply with the applicable regulations."[10] Vessel Traffic Services (VTS) and Vessel Traffic Information Service (VTIS), whether run by the Coast Guard or by other public or private agencies, use radio, radar, and visual information to monitor traffic. They gather and promulgate real time vessel traffic information, and broadcast traffic advisories and summaries. These are useful aids, but it is well to bear in mind that they in no way relieve the ship of responsibility for its own movements.

Shiphandling in a Narrow Channel

Transiting a narrow channel calls for particular skill on the part of the shiphandler. Knowing when to start a turn, how much rudder to use, and how to conn during the turn are skills that come with practice. The available shiphandling simulators are most helpful in providing the necessary practice in a safe environment. There are a variety of forces that come into play when conforming to a narrow channel. These include reduced responsiveness to the rudder in shallow water, bank suction, bank cushion, current, and wind.

The most basic skill in handling a ship in a narrow channel is how to handle turning points in the channel. There are two elements to a turn: timing the turn correctly and using the appropriate amount of rudder. The available shiphandling simulators are an excellent place to practice these skills to the point at which they become second nature. The turn should begin just before the ship's pivot point is at the turning point. In most cases a turn that is a little too early is easier to recover from than is one that is too late. It is not a good idea to plan for a full rudder turn unless it is a necessity, since that makes it difficult to tighten the turn if you have misjudged.

If you have turned a little too early it is usually possible to recover by easing the rudder, although it is usually not a good idea to stop the turn entirely. If you have turned too late, and an increase in rudder is not enough to tighten your turn, engines can be used. In a twin-screw vessel, backing the screw on the inside of the turn will tighten the turn. In a single-screw vessel, a short increase in an ahead bell to throw a wash against the rudder will usually tighten the turn.[11] This is another reason for proceeding at moderate speed in a channel: it provides a margin that lets you increase engine speed temporarily to improve rudder responsiveness, without gaining excessive ship speed. At very low speeds, bow thrusters or auxiliary propulsion units may be useful, but at normal channel speeds they are usually not effective.

Conning through Tight Passages

It is not uncommon to have to pass through especially narrow places in the channel, such as between bridge spans. If there is no wind or current, or if the set is parallel to the channel, safe passage is just a matter of staying in the center of the channel. If a center channel range is available it simplifies the task. If not, the best way to negotiate a narrow passage is for the conning officer to place him/herself on the centerline of the ship behind the centerline compass repeater. As the ship approaches the constriction the rate of relative bearing change of both sides of the channel should be the same. This can be observed visually by watching the apparent movement of the port and starboard obstructions against their backgrounds. The obstruction to starboard should be moving to the right against its background at the same rate as the obstruction to port is moving to the left. If these are balanced, the ship will pass midway between the two sides.

If the wind or current is setting the ship athwart the channel the passage can become more difficult. With a current from the side it is necessary to adjust the ship's course into the current (crab) to maintain the required track over the ground. This means that the ship is cocked relative to the channel course, effectively increasing its width (see fig. 7–2). If the opening is wide enough to accommodate this increase in effective width, then there is no problem, although it is well to stay on the up current side of the opening. If the increase in effective width is too wide to pass safely through the opening, then two alternatives are available. One is to go faster, which will reduce the angle needed to maintain track, proportionately to the increase in speed. It also increases the potential damage if something goes wrong. The second alternative, if water depths on the far side of the obstruction permit, is to curl

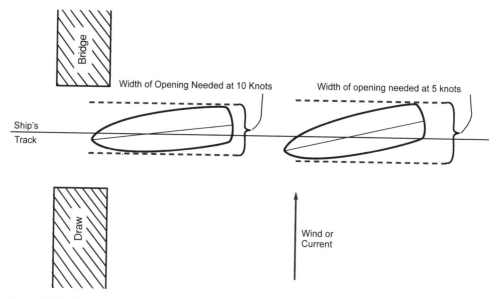

Figure 7–2. Cross channel set in a narrow passage.

the ship through the opening. To do this, the bow is aimed toward the up current side of the opening, and as the ship passes through the narrows and is set down by the current, the rudder is used to swing the stern away from the down current obstruction (see fig. 7–3). This alternative is available only if navigational considerations permit adequate room on the far side of the obstruction to swing the ship back into the channel.

Current

Current is common in a narrow channel, whether tidal or river generated, and affects the handling of the ship. A simplifying factor is that the current usually flows along the axis of the channel, avoiding the difficulties of a beam current. A current flowing in the direction the ship is headed is termed a "fair" or "following" current. One opposing her is a "head current." A head current is the safer, since it reduces the speed of the ship over the ground while maintaining sufficient speed through the water for maneuverability.

With a head current the best place to be is in the center of the channel. It is important not to let the ship get cocked with the current strongly on one bow. This can make it impossible to turn back into the current, and will push her toward the bank. Should this happen, two alternatives are available to sal-

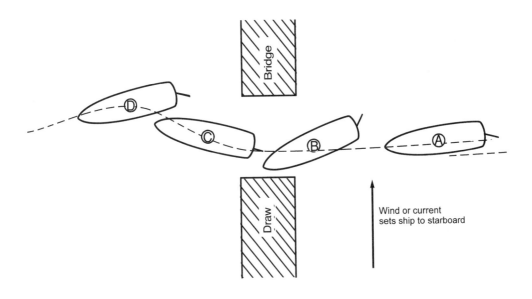

A. Bow close to upcurrent side of opening
B. As bow passes through opening, rudder is used to swing stern up current
C. & D. As stern is clear rudder is used to return to center of track

Figure 7–3. Curling the ship through a narrow passage.

vage the situation. One is to back down until the ship is stopped relative to the water and let the current carry her downstream until the situation can be sorted out. The current will normally keep the ship at a constant distance from the bank. Another, and perhaps preferable, course is to back until way over the ground is lost, then drop an anchor at short stay with engines stopped. This will cause the bow to pivot nicely up current, permitting us to retrieve the anchor and continue upstream.

With a fair current the ship's responsiveness is reduced, since the current reduces speed through the water relative to speed over the ground. It also prevents our using the anchor should we get into trouble, since that would cause her to swing sideways to the current. Fortunately, in most case the current tends to carry the ship along the center of the channel, and helps in turns. It is generally considered that the safest means of transiting a narrow channel with a following current is to stay close to the middle of the channel, slightly favoring the outside of turns.

Passing between two ships in a bend of a narrow channel should be avoided. With ships on opposite courses the ship headed upstream, because it is the more controllable, should wait for a downstream ship to clear before proceeding through a bend.

When turning from a channel into a slip or pier you need to anticipate a change in the effect of the current on your ship. It will typically change from a head or fair current to a beam current. If not anticipated and compensated for this can ruin your approach.

Bank Effect and Bank Suction

Bank effect and bank suction are encountered only in close proximity to steep banks. The effect is more pronounced in shallow water and is greater for large vessels than for small. In most circumstances in which these phenomena are likely to be encountered, pilots with local knowledge will be embarked.

Bank effect, sometimes called bank cushion, refers to the tendency of a ship to be pushed away from a close by steep bank. It is caused by the buildup of water between the ship and the bank, forcing the bow away from the bank. Bank suction, on the other hand, is caused by the Venturi effect between the ship's side and the bank. This can be amplified by the action of the ship's screws, particularly in a twin-screw vessel, evacuating water between the ship and the bank.

The combined effect of bank cushion pushing the bow away from a steep bank and bank suction pulling the stern toward the same bank can cause a sudden sheer for a ship transiting a narrow steep-sided waterway such as a canal. This is one of the reasons for the extra responsibility carried by Panama Canal pilots.

Overtaking

It is sometimes necessary to overtake another vessel in a channel. Both International and Inland Rules of the Road provide that in an overtaking situation it is the responsibility of the overtaking vessel to remain clear of the vessel being overtaken. The Rules also set forth requirements for overtaking in narrow channels. Both International Rules and Inland Rules provide for signals from the overtaking vessel and agreement by the vessel being overtaken.[12] If the ship being overtaken does not agree with a signaled proposal to overtake, she must sound five or more short blasts.

If it is necessary to overtake in a narrow channel, it may be done with safety so long as the maneuver is done properly. After the ship to be overtaken has agreed to be passed, she should provide as much room on the proposed side as navigational circumstances permit and slow as much as steerageway considerations permit to reduce the time required to complete the maneuver. The overtaking ship should pass at moderate speed, minimizing the time alongside and giving the overtaken ship as much room as possible. As is the case during an underway replenishment, you should anticipate some attraction between the two ships as a result of Venturi effect. The closer the passage and the higher the speed, the greater will be the effect.

The reason the Rules of the Road require agreement from the ship being passed is that they are the ones most likely to encounter a problem during the maneuver. No ship should agree to be passed until they are comfortable with the conditions. It is always worth considering the merits of having the overtaking ship have the patience to wait for more room for passing, or perhaps not passing at all.

Voyage Planning

Whether it is called voyage planning, passage planning, or a navigation brief, careful advance planning for a transit is something that competent professional mariners have always done. What is changing is the degree of formality required. The International Maritime Organization Convention of Standards of Training Certification (STCW) requires a passage plan "to establish written navigational procedures for preplanning a ship's route from berth to berth and establishing how the team will monitor progress along that route."[13] While the Navy and Coast Guard are not bound by the STCW, it is clear that they are moving steadily toward bringing military training and qualifications more closely in line with international standards. The type commanders' requirement for a written navigation brief in advance of any transit closely parallels the IMO requirement for a voyage/passage plan but is laid out in more detail.[14]

The purpose of a voyage plan is to anticipate the operational and navigational factors that will affect the ship from the time she gets under way until she is again moored. It is designed to provide a systematic tool to make your ship's transit safer and better organized. It is not in concept, and should not be in practice, a paper drill the effort of which exceeds its value. The purpose of a voyage plan is to ensure that all of the navigational and operational factors that can affect the transit are taken into account. It helps to guard against overlooked hazards and one-person errors. It is the professional alternative to "winging it." Providing a systematic plan helps bring attention to departures from the plan and facilitates prompt and correct adjustment.

It is the responsibility of the commanding officer to determine how detailed a voyage plan needs to be. Routine transits in and out of home port do not need the same level of detail as entry into an unfamiliar port, but a written plan is still required. Before entering restricted waters it is the responsibility of the navigator to prepare a navigation brief as a plan for safe and prudent passage, including piloting. The plan is to be reviewed by the executive officer and approved by the commanding officer.[15]

There are four stages to a voyage plan: appraisal, planning, conferring, and execution/monitoring. Appraisal involves collecting all information related to the voyage, identifying risks along the proposed track, anticipating potential problems. Planning establishes the routing, restrictions (such as speed restrictions), danger close points, danger bearings, and available navigation aids—everything that goes into a detailed "berth-to-berth" plan. The plan is developed to minimize the risks that were identified in the appraisal phase. The conferring stage involves everyone who has a role to play, from the commanding officer to the special sea detail helmsman in after steering.

In the Navy the conferring stage takes place at the navigation brief, details of which are laid out in the Type Commanders' Instruction 3530.4.[16] As is implied by the name navigation brief, the navigator has a large part of the responsibility for planning, but a number of others contribute. The navigator is responsible for briefing arrival and departure times, tides, currents, operational requirements, speed restrictions, conditions of readiness, and the tactical situation. If a geophysics officer is not embarked, the navigator also briefs weather, sunrise, sunset, moonrise, and moonset. It is the navigator's responsibility to identify the latest editions of charts, with corrections verified, and the GPS datum to be used with each chart, and to ensure that the charts and planned track are identical for CIC, the conning officer, and the navigator. The navigator is responsible for reporting the status of navigation equipment, including compass or repeater errors, any down equipment, backup systems, and ECDIS-N navigational systems. If tugs and/or pilot are to be used, the navigator's brief will include pick up and drop off points and times and the communications to be used.

The navigator, with help as needed from the operations officer, covers special considerations and events. These include anticipated honors and ceremonies, flag officer movements, visitors, helicopter operations, boats in the water, status of the accommodation ladder, any hot areas along the track, uniforms, and the watch bill. One of the most important responsibilities of the navigator in preparing the navigation brief is risk assessment. The risks to be considered include collision, grounding, navigation equipment malfunction, communications failure, man overboard, a breakdown in Bridge Resource Management, a steering or propulsion casualty, and reduced visibility. The assigned officer of the deck briefs the response for each of the potential emergencies.

The engineering officer (or reactor officer) briefs the status and configuration of the engineering plant, including the ship's degaussing system. Any limiting casualties and their impact are reported.

The scheduled conning officer is responsible for briefing the details of the track as laid down in coordination with the navigator. This includes courses, turn and danger bearings, ranges, shoal water, channel and turning basin depths, visual and radar NAV points, and the pier heading. If a vessel traffic separation scheme exists, this should be briefed, including check in/check out points. If anchoring, the conning officer should describe the anchorage, the type of bottom, head and drop bearings, and the amount of anchor chain required. The first lieutenant reports which anchor is to be the ready anchor, the scope of chain available, the status of windlass and winches, mooring lines in use, the procedures involved if mooring to a buoy, and any other pertinent details, as, for example, walking the anchor out if anchoring in deep water or dipping the anchor if going alongside another ship.

The navigation brief is prepared in written form and forwarded by the navigator to the executive officer for review and signature, then forwarded to the commanding officer for approval and signature. The navigator is to maintain a file copy for not less than six months. Properly prepared and executed, the plan minimizes the hazards of the transit.

Transiting the channel in a U.S. Navy or Coast Guard ship brings together all the skills, knowledge, and experiences of navigation, seamanship, and shiphandling. Navigation includes precise piloting using visual bearings, GPS, ECDIS, and radar. The use of anchors and mooring lines and considerations of weather, wind, and current, as well as maneuvering in close proximity to other ships are all parts of seamanship. Shiphandling requires knowledge of ship's characteristics, understanding of hydrodynamics, and knowing how to control forces to position her as desired.

In the performance of this complex evolution, the conning officer receives support from all of the ship's resources but still must rely on his/her own judgment, skill, and knowledge. Transiting the channel, as in all shiphandling, is both a science and an art, employing all of the tools of the mariner's profession.

8

TUGS AND PILOTS

————◄○►————

It was 0745 on a blustery October morning on the Chesapeake Bay. Wind was from the northwest at twenty-five knots, gusting to thirty-five. *Oak Hill* (LSD 51) was headed for a port visit to Baltimore. Special sea and anchor detail was set. *Oak Hill*'s skipper, Cdr. William Hastings, was familiar with the channel from Norfolk as far as Annapolis, so the ship had requested that the pilot join them there for the transit from the Bay Bridge to Baltimore. *Oak Hill* was flying the "Golf" flag, indicating that a pilot was required.

"Power boat flying Hotel flag at 340 relative, three miles," announced the port lookout.

The officer of the deck, Lt. Ray Kinnear, lowered his binoculars, acknowledged the lookout's report, and reported to Commander Hastings, "Pilot boat in sight off the port bow, Sir."

"Very well," said the skipper. "They've asked us to slow to three knots. Do so, and tell him by VHF that we are preparing to take the pilot on board at the starboard accommodation ladder. That should give him a decent lee."

After slowing the ship and communicating with the pilot boat, Lieutenant Kinnear instructed the junior officer of the watch to go below to the starboard accommodation ladder to welcome the pilot on board and escort him to the bridge.

Commander Hastings said, "We have room in the channel to sweep a lee as the boat makes its approach, so let the pilot boat know what you are doing, then put your rudder right full as he makes his approach."

"Pilot boat coming alongside," announced the telephone talker a few minutes later. Then, "Pilot on board. Pilot boat is clear."

Lieutenant Kinnear ordered, "Shift your rudder, come left to course 020, all engines ahead two-thirds, indicate turns for ten knots." To the signal bridge: "Haul down Golf. Close up Hotel." Then to the captain, he said, "I am returning to the channel course, Sir."

"Navigator concurs," announced the navigator.

Moments later the JOOW reappeared on the bridge, this time with a tall weather-beaten civilian in tow, introducing him to the skipper: "Captain Hastings, this is Captain Neilsen."

"Welcome to *Oak Hill,* Captain Neilsen. How would you like your coffee?"

"With just a couple of teaspoons of sugar, thank you."

The boatswain's mate of the watch handed a steaming mug of coffee to the pilot and said, "The galley made some fresh doughnuts this morning, Captain. Would you like one?"

"Thanks, I would," said Captain Neilsen, "but perhaps a little later after I get settled in."

"Here is a summary of *Oak Hill*'s characteristics," said Commander Hastings, handing a laminated card to the pilot. "We have no casualties that will affect our transit. With this wind, though, we are going to have to exercise some care. The ship has a lot of sail area, mostly forward. We've been having to crab several degrees coming up the channel."

"Yes, that's what I would expect," said the pilot. "We won't have any significant current at your berth, but we will have to pay attention to the wind. Although ordinarily I'd be happy handling a twin-screw ship like this with only one tug, with the wind I've asked for a second one to join us. They will be rendezvousing with us at Brewerton Angle. How would you like to conduct the pilotage? What do you want from me?"

"Two tugs sounds good to me," said the skipper. "If it suits you, I'd like Lieutenant Kinnear to keep the conn until the tugs join up, then we'll turn it over to you. In the meantime, I'll count on you to back us up with advice as needed."

"Whatever you want, skipper. Traffic is going to be pretty heavy this morning, particularly tugs with tows. I'll volunteer to use the VHF to keep us sorted out with the traffic, since I already know a lot of these guys."

"I'll appreciate that," said the skipper. "If it is all right with you, once you take the conn I'd like to have a couple of our officers who are working on their shiphandling skills stand by as observers, if you'd be willing to explain what you are doing, and why. As a matter of fact, I'm likely to learn a few things as well."

"I'd be glad to," responded the pilot. "I'm always glad of an attentive audience."

As *Oak Hill* neared Brewerton Angle, the tugs were seen waiting. Captain Nielsen used VHF to direct them to follow. As they approached Fort McHenry, he directed one to make up to the port bow with a headline, the other with a headline on the port quarter. To the observers he explained, "We will be going starboard side to, and the wind will be setting us on to the pier. There is no significant current. By making up the tugs this way I can make our approach with some way on, then use the tugs to both help slow us and to control the speed with which we will be set onto the pier. I plan to make our approach parallel to the pier about one hundred feet out and let the wind move us down onto our berth." Captain Neilsen made his approach to the pier at five knots, then ordered all engines back one-third as the bow passed the end of the pier. This was followed by orders to both tugs to back half and work out to 90 degrees. As way came off the ship he ordered, "All engines stop." As the bow started to move toward the pier he ordered the forward tug to back full, bringing *Oak Hill* back parallel to the pier. All lines were ordered over. With a final order to both tugs to back easy, *Oak Hill* settled gently on to the camels. "Hold all lines." "Double up all lines." He then released the tugs, each of which acknowledged with a prolonged and two short blasts on their whistles.

Commander Hastings smiled at the pilot. "Captain, you make it look too easy," he said. "I'm afraid we may be making these young gentlemen think it is simple, when I know better."

"Well, when you've been doing it for almost forty years, it does get easier—and all we had to contend with today was a little wind," Captain Neilsen replied. "The fact that your ship is more powerful and responsive than most of the ships I handle is a treat." He paused then said, "Oh, and if that doughnut is still available, I'd be happy to take it with me."

As the size and tonnage of naval vessels have increased so has the use of tugs and pilots. A World War II destroyer of the *Fletcher*, *Sumner*, or *Gearing* class had a full load displacement of less than 3,500 long tons and an overall length of 376 to 391 feet. By comparison, a modern *Arleigh Burke*–class guided missile destroyer has a full load displacement in excess of 8,300 tons and an overall length of 504 to 509 feet. The World War II destroyers had less sail area than their more modern counterparts, and their sonar domes were tucked beneath the keel, making it safe to work the bow in against the pier. Surprisingly, the World War II destroyer had a substantially better power-to-weight ratio than that of the modern *Arleigh Burke*.[1] All other things being equal, a smaller, lighter ship with a higher power-to-weight ratio is easier to handle and more forgiving than a larger vessel. The commanding officer of a World War II destroyer could take legitimate pride in handling his ship in virtually all circumstances without the assistance of either tugs or pilots. That is no longer the case. Prudence now sometimes dictates that even the most skilled naval shiphandler should use the available assistance.

There are, however, very good reasons not to become completely dependent upon tugs and pilots. Warships must be able to operate on short notice under all conditions, whether or not tugs and pilots are available. It is in emergency conditions that they are least likely to be readily available. Although most pilots are highly competent and experienced seamen, from time to time we may encounter one who is not so professional. In ports visited only infrequently by men of war, a pilot who is highly skilled at handling merchant ships may have difficulty in quickly adapting to the much greater power and responsiveness of some naval vessels. In foreign ports, language difficulties can exist, and in some cases pilots may not even be available. For these and other reasons, a competent naval shiphandler needs to understand piloting skills and the use of tugs.

Relations with the Pilot

The presence of a pilot on board in no way relieves the commanding officer of his total responsibility for the safety of the ship. The only exceptions are transit through the Panama Canal, entrance into a dry dock (in which case the docking officer assumes responsibility as the entering part of the ship passes over the sill), and certain cases in German waters. With these exceptions the pilot's legal role is that of an advisor to the ship's captain, whether or not he actually assumes the conn.

The navigation instruction issued jointly by commander Naval Air Forces and commander Naval Surface Forces states:[2]

> Commanding Officers must be especially dutiful while operating in foreign waters to maintain the safety of the ship when evaluating the

recommendations of an embarked pilot, especially when the pilot recommends deviating from the planned track to avoid shipping. Pilots, as advisors to the Commanding Officer and the Navigation Team should be familiarized with ship's characteristics and planned navigation track prior to beginning the proposed transit.[3]

The instruction goes on to provide a checklist of items for discussion with the pilot, as follows:

1. Maneuvering characteristics of the ship and lowest depth projection;
2. Allowable deviation from track;
3. Unpublished hazards to navigation;
4. Bridge-to-Bridge radio communications;
5. Ship-specific piloting and conning procedures;
6. Use of tugs;
7. Material casualties that may affect maneuverability of the ship;
8. Material condition of ship (oil leaks, steering system, etc.);
9. Safe speed for all legs of proposed transit;
10. Status of ECDIS-N system and correction status of electronic charts.[4]

It is useful, particularly for foreign pilots, for the ship to provide a laminated diagram of ship's characteristics. This should include navigation draft, configuration of screws and rudders, capstans and ground tackle, location of sonar dome, overhangs, and so on. In the interest of the best working relationship, however, it may be best to first hand the pilot a hot cup of coffee, and perhaps even a sandwich. Make sure that the niceties of greeting and briefing do not distract attention from the vital tasks of keeping the ship in safe water, avoiding harbor traffic, and lining up for the channel entrance. Do not be too hasty in giving the pilot the conn. Give him time to get dark adapted, oriented, and thoroughly briefed. Only then should he be given the conn, if that is the intent.

Perhaps the best use that can be made of the highly honed skills of a good pilot, besides actually conning the ship is, if he is willing, to use him as an instructor while he is on board. This may take some diplomacy, since the inclination of most pilots is to take the conn, transit the channel, and berth the ship. It is much more advantageous to the ship, however, if he can be prevailed upon to explain to a selected small group of fledgling shiphandlers what he is doing, and why. The emphasis here is upon small, since as stated elsewhere, nothing should be permitted to interfere with having a quiet and orderly bridge. The commanding officer himself should not be reluctant to place himself in the learning group if his opportunities to handle tugs have been limited.

In other cases, the commanding officer may choose to use the pilot in a purely advisory role. If the professional competence of the pilot is unknown, or a language difficulty exists, this can be the best solution. One way to do this that works well is for the pilot to conn with the commanding officer relaying his orders to the helm and engines. This lets the ship's captain exercise his own judgment while still taking advantage of the pilot's local knowledge. Another alternative, particularly useful when, as sometimes happens, the pilot does not use standard commands, is to have a well-qualified ship's officer translate and relay the pilot's commands. It is probably a good idea, however, particularly where language difficulties may exist, to have the pilot handle all voice communications with the tugs.

Picking up the Pilot

When entering port the pilot will almost always be picked up from a pilot boat. The receiving ship establishes communication with the pilot boat, selects a course that provides a good lee for the pilot boat's approach alongside, and proceeds at slow speed. The course selected should be one that minimizes roll, as well as providing a good lee. In most circumstances, three to four knots is a good speed, allowing the pilot boat a useful speed advantage for his approach. In some cases, particularly if a relatively high power pilot boat is being used, the pilot may request a higher speed. This kind of transfer can be hazardous, and it is the obligation of the receiving ship to make it as safe as possible. Even seas that seem fairly calm as viewed from the bridge can be rough at the boat's level. Swinging the stern away from pilot boat just before the pilot boards can often make the water calmer by "sweeping a lee" (see fig. 8–1). This maneuver can create water flow toward the ship, sucking the boat alongside, so it is important to warn the pilot boat of your intentions.

The ladder for the pilot should be as safe as it can be made to be. Naval vessels have not always done this as well as they should. The ladder should be properly rigged and inspected before the arrival of the pilot boat. Solid communications are necessary between the bridge and the boarding station. It is best for bridge, pilot boat, and boarding station to all be on the same radio channel. Sailors should be standing by to assist as necessary, and the receiving location must be properly lighted. Make sure the lights are aimed so that they

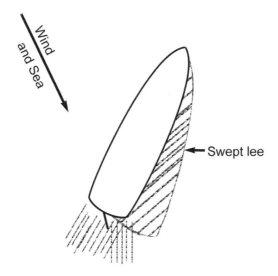

Figure 8–1. Sweeping a lee.

will not blind the boat operator or the pilot while coming aboard. A lighted life ring with a line attached should be on station. A heaving line should be lowered to pick up the pilot's bag. The boarding ladder itself should be rigged so that the bottom rung is above the deck of the pilot boat. A longer ladder can be damaged by being caught between the boat and ship, and for the safety of the pilot it is important that once on the ladder he is well clear of the boat. Stanchions should be provided as handholds at the top of the ladder.

Capt. Malcolm Armstrong has set forth the characteristics of a good pilot ladder:

1. The rungs should be of one piece hardwood;

2. The bottom four rungs should be of reinforced hard rubber;

3. All of the rungs should have a nonskid surface;

4. The rungs should be at least nineteen inches long, four inches deep, and one inch thick;

5. The rungs should be painted a high-visibility color, preferably white or international orange, and should be non-skid, not gloss;

6. The line used should be low stretch—manila or low-stretch dacron;

7. The clear space on each rung should be from sixteen to nineteen inches;

8. The rungs should be spaced at equal distances, from twelve to fifteen inches apart;

9. If the ladder has more than nine rungs, spreaders are required to keep it from twisting. Spreaders should be at least six feet long, placed behind and in line with rungs, and spaced at intervals of no more than nine rungs.[5] If the ship to be boarded has high sides (more than about thirty feet), an accommodation ladder should be rigged in addition to the pilot ladder to provide more secure footing.

When we are headed to sea and the pilot leaves us, many of the same considerations apply. The ship slows, and maneuvers to create a lee for the pilot boat to come alongside. There will often be other traffic nearby: notify them of your intended maneuver to avoid any possible confusion. Considerations for manning the boarding station and rigging the pilot's ladder are the same as for bringing the pilot on board. Make sure the pilot boat is well clear before gaining speed and maneuvering to resume track.

Using Tugs

A skilled shiphandler needs to know how to use tugs. As with any other skill the effective use of tugs requires practice. The required knowledge includes the characteristics of the tugs themselves, where to place them, how to make them up and how to communicate.

Tugs are a diverse lot. Most of them are designed for towing. Those intended for shipwork will generally have small wheelhouses and masts that can be lowered easily to permit them to work under ship's overhangs. Since it is not unusual for a tug designed for towing to be called into service for shipwork, however, this can affect where the tug may safely be placed on the ship.

The larger the ship and the stronger the current and/or wind, the more important the power of the tug becomes. Because at low speeds the ship can be considered to be operating in a frictionless medium relatively little power is necessary to move it. When it is necessary to oppose wind and current, however, considerable power may be required. The amount of power needed varies with the square of the velocity of the wind and current, with their direction, and with the tonnage and sail area of the ship. It can also vary with the configuration and makeup of the tugs. The horsepower that a harbor tug can deliver varies widely, with a normal range from 750 to 5,000 horsepower or more. The horsepower rating is what the tug can develop when going ahead. Most tugs deliver considerably less power when backing than when going ahead. Because of the number of variables involved, there is no simple rule of thumb to use in determining how much tug power is enough under difficult conditions. Because of his familiarity with local conditions and the local tug fleet, the pilot's advice is useful. In case of doubt, a bit more power than needed is always better than not enough.

At one time most large Navy ports had Navy tugs assigned, and a tour as a tug master was the pinnacle of the career of many chief boatswain mates. More recently Navy tug services have been outsourced to commercial contractors. For example, at this writing, the Moran Towing Corporation held the contract to provide harbor services to Navy ships in the Norfolk area. Moran has eight modern and capable tugs assigned to these duties under long-term charter, and can provide more on relatively short notice, if needed.

Single-Screw Tugs

Tugs vary widely in their ability to maneuver. Single-screw tugs are the least maneuverable, but many of them may still be found around the world, even though they are now basically old-fashioned technology (see fig. 8–2). In circumstances in which there is a possibility of the screw contacting the ship, as in working alongside a submarine, which is wider below the waterline than above, a single-screw tug may be preferred. The single-screw tug shares many of the handling characteristics of a single-screw ship, but exaggerated because of the relatively large diameter and pitch of the tug's propeller. The backing power of the tug is markedly less than its power when going ahead and the stern will have a pronounced walk when backing. As with a ship, the direction depends on shaft rotation. Most tugs have a right-hand rotating screw and thus will back to port. Like a single-screw ship, the tug can be turned by backing and filling much more readily in one direction than the other. With a right-hand rotating screw, the tug can be turned readily to starboard, but much more slowly to port. To maintain its position when alongside, the single-screw tug may need to exert pressure against the ship. Depending on circumstances, this pressure may be unwanted. Because of their relatively restricted maneuverability, single-screw tugs are more dependent upon the use of lines to hold them in place. These limitations are normally not serious, however, and a properly made-up and used single-screw tug can be of great assistance to the ship.

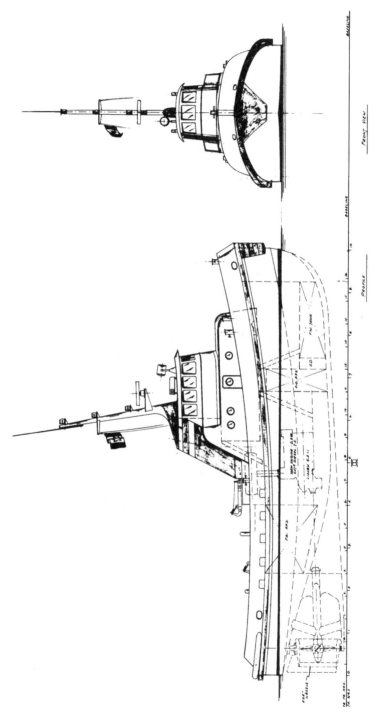

Figure 8–2. Single-screw tug with steerable Kort nozzle and CP propeller. *Robert Allan Ltd.*

Twin-Screw Tugs

Twin-screw tugs are more maneuverable, usually more powerful, and have an edge in reliability through the redundancy of twin engines. Favored since the mid-sixties for the majority of coastal and some harbor tugs, the twin-screw configuration usually comprises open or nozzled propellers, each with single or multiple rudders behind. The vessel shown in figure 8-3 is fitted with twin nozzled propellers and twin high-aspect-ratio rudders. The additional control provided by twin screws permits the tug to remain in position with less dependence on lines. Because the shafts rotate in opposite directions, the tug can back straight. When both engines are backing a twin-screw tug can generate about two-thirds of the bollard pull it has when going ahead. Through the use of opposed engines the twin-screw tug can maintain a 90-degree angle alongside even when the ship is moving at up to two or three knots. It cannot, however, generate nearly as much power ahead or astern in this circumstance because of the need to have an opposed engine. The use of a stern line will aid the tug in maintaining position when the ship has sternway, but caution must be exercised since in this position the tug is vulnerable to tripping.

Tugs with Directed Thrust

Many modern tugs use one or another means of directing their thrust. This makes them even more maneuverable than a twin-screw tug and permits them to use their full power in circumstances that would require a twin-screw tug to operate with opposed engines. The use of directed thrust also permits the tug to steer effectively when going astern. Some of the types of directed-thrust tugs are the Kort nozzle, flanking rudders, and a variety of propeller-steered designs.

The Kort nozzle is a circular shroud enclosing the tug's propeller, enhancing its efficiency. It is usually single screw and can be either fixed or steerable. The fixed Kort nozzle increases the tug's thrust ahead. It is most effective when combined with flanking rudders, as discussed below. The steerable Kort nozzle directs the thrust from the propeller, as controlled by the tug's steering gear. The screw remains fixed, but the nozzle angle changes. Although mechanically complex, it makes for a very controllable tug.

Flanking rudders can be used either with Kort nozzles or conventional propellers. The flanking rudder design puts rudders both ahead and behind

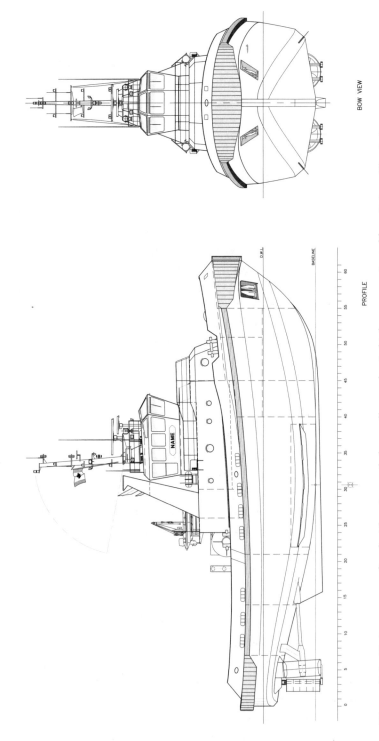

BOW VIEW

PROFILE

D.W.L.

BASELINE

NAME

Figure 8-3. Twin-screw tug fitted with twin nozzled propellers and twin high-aspect-ratio rudders. *Robert Allan Ltd.*

the propeller or propellers. Thus whether the tug is going ahead or backing there is direct thrust from the propellers onto a rudder. This gives the tug excellent maneuverability, allows it to back under good control, and enhances its ability to maintain working position without depending on lines.

Propeller-steered tugs are similar to the steerable Kort nozzle, but without the shroud. They are steered directly by their propellers, rather than by rudders. They have excellent maneuverability, the ability to maintain position without depending on lines or compromising their backing power. Propeller steered tugs can be either single or twin screw. They are either conventional pusher tugs with the propellers located in the traditional aft location or tractor tugs with the propellers located toward the bow. Either configuration works well, and both pusher and tractor versions have almost as much power astern as ahead. There are a number of variants of propeller-steered tugs. Among them are the Voith-Schneider, the Schottle drives, the Harbormaster, and the Z-drive (see figs. 8–4, 8–5, and 8–6). The differences in design do not translate to significant operational differences, and can in most cases be ignored by the shiphandler.

To properly use tugs, the shiphandler needs to know the characteristics of the assigned tugs. The configuration of the tug will determine how to make best use of it. Yet even a low-powered, single-screw tug can be of great assistance when properly used. When there is surplus tug power the type of tug is less important. When power is marginal, the type becomes very significant.

Tug Equipment

For a tug to work alongside a ship, it must have fenders designed to prevent damage to the ship or to itself. The most important of these fenders is the bow fender, which when the tug is pushing transfers the power of the tug's engine to the side of the ship. To prevent damage to the relatively thin shell plating of the ship it is necessary that the tug's bow fender be designed to distribute the pressure. Heavy synthetic rubber extrusions have largely replaced the older rope bow fenders. The rubber extrusions are better at preventing damage, but their generally lower coefficient of friction makes it more difficult for the tug to maintain its pushing position without lines. Most tug bow fenders are mounted only above the waterline. For working with submarines, however, a tug must have bow fenders that extend well below the waterline to the contact point on the submarine's underwater hull.

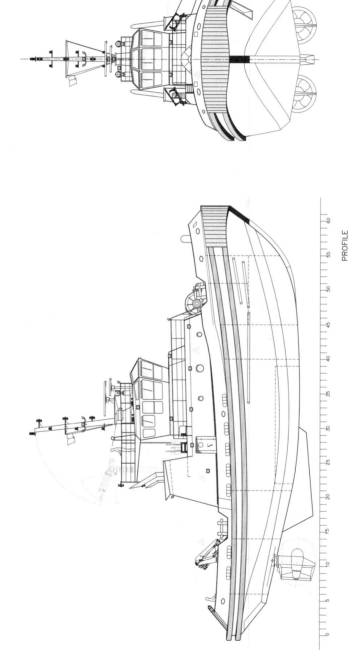

BOW VIEW

PROFILE

Figure 8–4. Tug with azimuthing stern drive (ASD) configuration. *Robert Allan Ltd.*

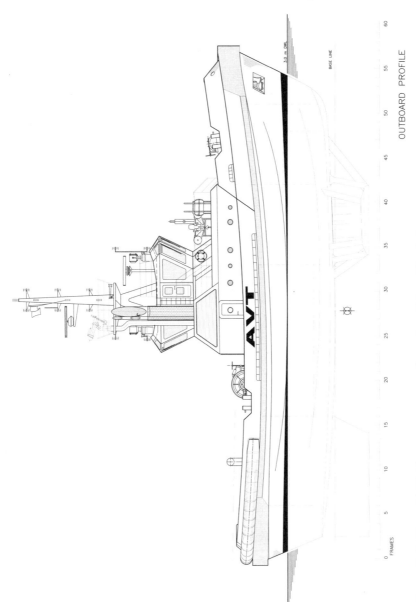

Figure 8-5. Tug with Voith water tractor configuration. *Robert Allan Ltd.*

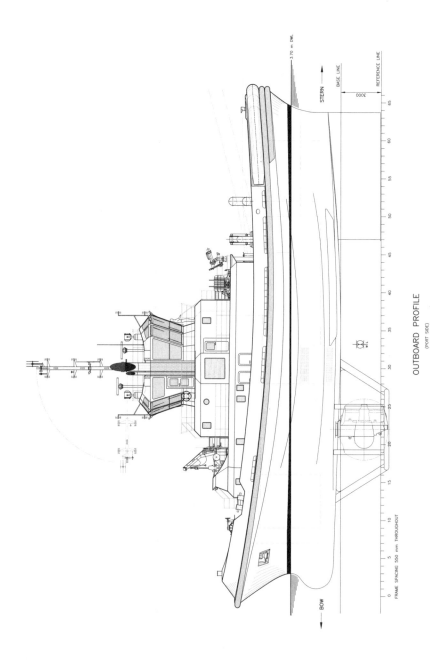

Figure 8–6. Tug with Z-drive tractor system. *Robert Allan Ltd.*

Tugs are also normally well equipped with side fenders. Some use purpose built fenders, but for reasons of cost worn out truck tires are most often used. The effectiveness of the tire fender can be improved by wrapping them with used line to improve their durability and bulk. This has the collateral advantage of eliminating the tendency of the tires to mark the ship's sides. Stern fenders are seen more rarely, primarily on tractor tugs designed to push against the side stern first. These fenders are usually synthetic rubber extrusions similar to those used on the bow.

Working lines are another important part of a tug's equipment, and the tug will usually provide its own lines to the ship. These lines will normally be of synthetic fiber, such as polyester, polypropylene, or polyethylene, but *not* nylon. Nylon, although strong, stretches too much to provide a solid makeup for the tug and should it part under strain is likely to be lethal. The lines provided by the tug may have a wire rope pendant with an eye large enough to drop over the ship's bitts. This is done to eliminate the chafing of the softer line as it goes through the ship's chocks. As the wire rope ages, it develops "fishhooks" from broken strands, and it is important that the line handlers be provided with heavy gloves to avoid injury. The wire rope can also abrade the interior of the ship's chocks, making them rough and likely to damage any soft lines led through the chock. Chocks should be inspected regularly and smoothed if necessary. In making up the tug, it is useful to have messenger lines led out through the ship's chocks to be used in picking up the tug's lines.

Positioning Tugs

In positioning tugs for use, both safety and effectiveness need to be taken into account. Single-screw tugs in particular can have a difficult time getting away from alongside a vessel, especially when they are port side to, because of their strong sternwalk to port when backing. Some ship's characteristics can pose hazards for tugs. Submarines are wider under water than at the surface, and the tug's fenders provide little useful protection if the tug contacts the submarine other than with a purpose-designed underwater bow fender. Other naval vessels, particularly aircraft carriers, have side projections that can contact the tug's mast or superstructure. Multiple-screw vessels can have propellers that extend beyond the sides of the ship. Some ships have hull flares that prevent tugs from being able to work closely alongside. Nets around helicopter decks are especially low and vulnerable to damage. In some ships, overboard discharges may be located such that they could discharge on the tug's deck. All of these characteristics must be taken into account in deciding where to locate and work tugs.

Communications with Tugs

At one time the principal means of communication between ships and tugs were hand and sound signals. As ships have increased in size and radio communications have become more convenient and reliable, VHF radio has become the communications medium of choice. Unfortunately, there is as yet no universally adopted standardization of verbal commands to tugs by VHF. An experienced pilot, Capt. Vic Schisler, has proposed a set of suggested standard commands. More detailed information, including advice on the use of tugs may be found in Appendix B, but some of the basic principles are described below:

1. Always give tug name before giving commands. This will alert the tug captain that what she/he hears next will apply to her/his tug.

2. Do not use given names of tug operators, as this may leave the ship's bridge out of the information loop and there may be more than one "Bob" working tugs at that time.

3. Use power references of EASY (for 1/3), HALF (for 2/3), or FULL (for 100 percent).

4. Reference tug commands to own ship. Specify the direction relative to the assisted vessel; for example, "*Guard* pull easy to port."

5. When asking for a push or pull, include the direction that the force is to be applied, port, starboard, forward, or aft; for example, "Pull easy to starboard," "Pull easy to starboard 45 degrees forward (aft)."

6. The tug operator should attempt to maintain the last angle requested by the pilot until changed by another command, that is, the pilot may request that you "stop at 90" (degrees) or "stop and drag on your line."

7. The majority of these commands apply to conventional tugs as well as "tractor," "combi-tug," and "ASD" (azimuthing stern drive) type tugs.

It is often more convenient for the tug to acknowledge an order by a whistle toot, and it is always prudent to have sound signals as a backup if radio communication fails.[6]

Sound signals are not universal, but there are some that are in relatively common use. If the tug is going ahead or astern, one short blast is the signal to stop. If the tug is stopped, one short blast is the signal to go ahead. Two short blasts tells the tug to back. Three very short blasts tells the tug to increase to full speed in the direction it is already going. One prolonged blast signals the tug to go more slowly in the direction it is already going. One prolonged blast followed by two short blasts signals the tug to cast off or to shift positions. (A short blast lasts one second; a blast, two to three seconds; and a prolonged blast, four to five seconds.)

Signals to Tugs

Sound signals: Short blast = one second *
                 Blast = two to three seconds—
                 Prolonged blast = four to five seconds—

Signal	Meaning
–	From stop to half-speed ahead
–	From half-speed ahead to stop
****	From half-speed ahead to full-speed ahead
–	From full-speed ahead to half-speed ahead
—	From stop to half-speed astern
****	From half-speed astern to full-speed astern
–	From half-speed or full-speed astern to stop
——**	Cast off, stand clear

The tug will acknowledge each order with a short whistle blast, except for the backing signal, which is acknowledged with two short blasts and the cast-off signal which is acknowledged by repeating the one prolonged and two short blast signal.

Use of Tugs in Shiphandling

A tug can push or it can pull. Its ability to assist the shiphandler depends on being properly positioned and connected to the ship. To do this, it is necessary to think through the intended maneuvers in order to get optimum assistance from the tug or tugs. If there is a strong current or wind, optimum tug positioning will provide an offsetting vector. If the channel has tight bends, the tug can be placed to assist in turning the ship. A tug can be used to hold the ship against the pier while handling lines. Perhaps more than anything, the tug provides an extra margin of safety in the event of equipment casualty or the misjudgment of a maneuver.

A Single Tug

The applicable rules for employment of tugs vary from port to port, but a typical rule is that a warship can utilize a single tug without taking a pilot but is required to take a pilot if two or more tugs are to be used. For this reason, and because the use of a single tug illustrates most of the principles applicable to multiple tugs, we turn first to the use of a single tug.

Because the ship's own engines and rudders provide maneuverability to the ship's stern, we will most often want to use a single tug to control the bow. An exception would be a ship with a bow thruster or auxiliary propulsion units as found on the *Perry*-class FFG, although if wind and current conditions are adverse even they may need a tug. Since these ships can be handled in most circumstances without tugs, the focus here is on ships without these features.

One of the most flexible ways to use a single tug is to make it up on the bow facing aft, or on the quarter facing forward. This is called "on the hip," "hipped up," or a "power makeup," depending on local usage. The makeup on the bow facing aft is sometimes called a "Chinese power makeup." In each of these cases, the tug is made up with a minimum of three lines: a headline, a stern line, and a spring line. The tug should be made up "flat," with its keel parallel to the keel of the ship and on the side opposite to the pier. The ship needs to be making slight sternway while the tug is making up in a reverse power makeup. This configuration can be of particular use when backing into a narrow slip.

The use of a tug hipped up on the quarter is shown in figure 8–7. This makeup is most often used with a single-screw ship, since the effect is to provide essentially the maneuvering characteristics of a twin-screw ship. It is also sometimes used for a "dead stick" move, in which case it serves as a substitute for the ship's own screw and rudder.

The use of a tug hipped up on the bow headed aft is shown in figure 8–8. This is one of the most flexible and controllable ways in which to use a tug. The basic technique is to balance the thrust of the tug coming ahead against the ship's ahead bell. This provides very precise control in all planes. It also minimizes the need for frequent engine orders to the tug, since adjustments to the ship's engine provide precise fore and aft control. As can be seen in figure 8–9, by placing both the tug's and ship's rudders right the ship will rotate clockwise. If they are both placed left, the ship will rotate counterclockwise. The ship can be moved laterally to starboard by placing the ship's rudder to the left and the tug's rudder to the right. Opposite rudders will move the ship

Figure 8–7. Tug hipped up on the quarter (also termed a power makeup).

Figure 8–8. Tug hipped up on bow (also termed a reverse power makeup).

to port. As is evident, this gives excellent control. In adverse conditions, however, the amount of lateral power vector that can be achieved this way is less than if the tug were being used to push or pull.

The simplest way to use a tug is as a pusher, with no lines to the ship. Used this way, a tug can help the ship to pivot, or to move laterally, depending upon where it is placed. To move the ship laterally, the tug should be placed as close as possible to the ship's center of lateral resistance (CLR), which with no way on will be the center of the submerged portion of the ship's side. The further the tug is ahead or aft of the CLR the more rotating motion it will impart, and at the same time the ship's pivot point will move toward the end opposite to where the tug is placed. This effect becomes more complex when the ship is moving, as the pivot point moves forward when the ship is going ahead, and moves aft when she goes astern. This is illustrated in figure 8–10. Note that because of the movement of the pivot point, when the ship is going ahead a tug pushing on the port side amidships will rotate the ship counterclockwise. If she is dead in the water, a tug pushing amidships

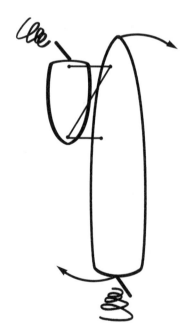

Figure 8–9. Rotating ship with tug in reverse power makeup. Ship and tug both ahead, balancing fore and aft vectors. Ship is rotated (or walked laterally) by use of rudders.

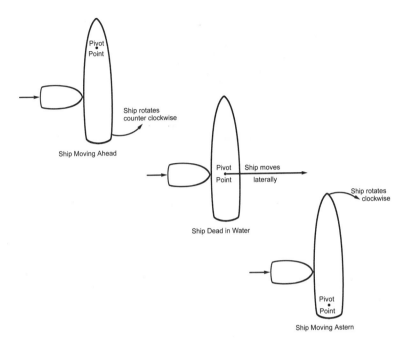

Figure 8–10. The moving pivot point.

will move the ship laterally with little rotation. If the ship is moving astern, that same tug pushing amidships will rotate the ship clockwise. Depending on its propulsion and steering system a tug may have difficulty maintaining a position at 90 degrees while the ship is moving ahead or astern. A quarter line from the tug can be used to keep the tug from falling off as the ship moves ahead, but this must be used with caution, since too much way on the ship could cause the tug to capsize.

A third way to make up a tug is with a single headline. The considerations for using the tug this way are much like those discussed above for the unsecured tug, except that with a headline the tug can either push or pull. If the tug is made up forward with a headline, it can ride safely alongside the ship at somewhat higher speeds than the two or three knots maximum otherwise desirable. In this position (see fig. 8–11) the tug, if backed, will both slow the ship and exert some rotating force on the ship. As the ship slows, the tug can work out to a 90-degree position to assist with a docking.

In the United States, tugs are not often used with towlines for docking, but are used that way much more often in European ports. An exception

Figure 8–11. Tug with single headline.

might be the use of a towline in the United States to move a ship out laterally from the pier when getting under way, but even in this case a headline is often preferred. (Note, however, that if power is a consideration, as in the case of a strong beam wind or current, that most tugs do not generate as much power when backing as when going ahead.)

The basic use of a towline is simple in concept. The tug pulls the ship in the direction you want to go. This can be straightforward if the tug is assisting the ship through a narrow waterway, getting under way from alongside, or turning the ship in a turning basin. It is not of much use in moving a ship laterally toward a pier. The tow line to the tug needs to be long enough to minimize the effect of tug prop wash on the hull of the assisted vessel.

The same considerations on location of the tug and the effect of the tug and the ship's motion on the pivot point apply with the use of a towline. Towlines can be more hazardous, however, because the towline can capsize the tug if the ship gathers excess way before the tug is released. European tugs are less subject to this hazard because they generally use a towing hook set well forward from the stern. The hook can be tripped from on board the tug if a hazardous situation develops, and the forward position of the hook permits substantial maneuverability even when the tow line is under strain. The same situation can negatively affect the maneuverability of an American tug with the tow point located further aft.[7]

Figure 8–12. Tug with single headline assists USS *Antietam* (CG54) in getting under way from her berth in San Diego. *U.S. Navy photo*

A tug can be very useful in mooring to a buoy. Make the tug up with a single headline, and it can keep the ship's head lined up with the buoy by pushing or pulling as necessary. The ship's own engines are used to maintain fore and aft position on the buoy. In many cases, particularly with larger vessels, the buoy will not be readily visible from the bridge of the ship when in mooring position. If this is the case, it is desirable to move the conn to the forecastle.

Multiple Tugs

Multiple tugs are required when the size of the ship or the strength of the wind and current dictate. The largest warships, such as CVNs and LHAs, always use multiple tugs and a pilot. Although with smaller ships dead stick moves can sometimes be accomplished with a single tug, it is desirable to use two or more if available.

From a shiphandling point of view two tugs will permit us to carry out any desired evolution, so long as it is within the power limitations of the tugs. The mass and inertia of large ships and the effect of wind and current can increase the number required. When using multiple tugs with a large ship entering a harbor, it is useful to position them in symmetrical pairs. The first pair should be assigned to the bow, since the ship's engines and rudders provide some controllability to the stern. The next pair will be located at the ship's center of lateral resistance to assist in positioning the ship laterally. Additional tugs may be placed aft. As the ship approaches the pier the tugs on the pier side need to be cast off in plenty of time to get out from between the ship and the pier. If needed, they can then make up on the side of the ship opposite the pier.

If the channel into the harbor is narrow and has tight turns, it can be useful to place a tug with towline directly ahead, and another with a headline directly astern to help maintain the ship's alignment in the channel, and to tighten the ship's turning radius as necessary. The placement of a tug directly astern also permits control of the stern in case the assisted vessel suffers a propulsion or steering casualty.

Figure 8–13. Tugs assist USS *Carl Vinson* (CVN 70) as she moves pierside in Singapore. Note the way in which the carrier's overhang dictates where the tugs can safely work. *U.S. Navy photo*

If a ship is to be moved without power, tugs must provide motive force as well as lateral control. Probably the best makeup for this is to use two tugs in a power (hipped up) makeup on either quarter of the ship. This then permits the ship to be handled identically to a twin-screw ship, but with even more maneuverability because of the wider spacing between the tug's propellers.

It is well worth making time in the schedule to pay a visit to the tug masters in your home port, or to invite one or more of them to lunch on board and a seminar on the use of tugs. There are few better ways to learn what to do and what not to do, and establishing a relationship might even help to make your tug services even better.

9

UNDERWAY REPLENISHMENT

———◄◦►———

USS *Higgins* (DDG 76) was steady on course 185 degrees true and making thirteen knots as the officer of the deck, Lt. (jg) Ed Hart, stepped out of the sun and wind of the starboard bridge wing into the pilothouse just as the captain entered from the port side.

"Cap'n's on the bridge!" sang out the bos'n mate.

Ed saluted. "Morning Captain. I hold *Yukon* at 214 degrees true at three miles. She's on Romeo Corpen 005, thirteen knots with *Milius* alongside to port. *Elliot*'s in lifeguard one thousand yards astern. We're to take waiting station on the starboard quarter of *Yukon*. We're going our port side to her starboard side. Right now I'm steering 185 at thirteen knots, so we're on a reciprocal course to *Yukon*'s 005." (See fig. 9–1.)

The captain smiled as he returned the salute. "Seems like you've got everything under control, Ed. Whose turn is it for this UNREP?"

"Well," Ed said, glancing at the navigator and then at the junior officer of the deck. He smiled. "I was planning to do this one myself, Captain."

"Yourself? You mean you want to do it all? How about your team-mates?"

"Oh, I'll take all the help I can get, but I'd like to run the whole evolution this time. Join-up, approach, alongside, and clearing"

The captain thought a couple of seconds then said, "Okay, tell me how you're going to do this."

Ed described the process as he had planned, including, "I'm going to continue on this course and speed until *Yukon* bears 230, range

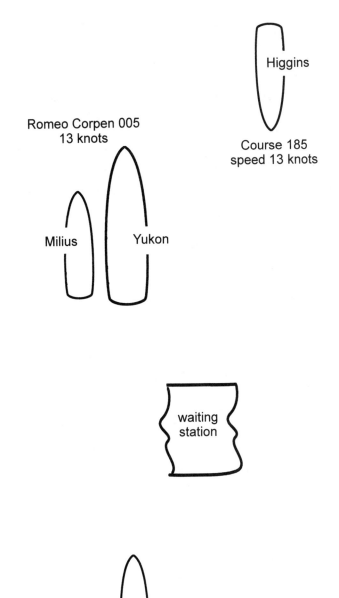

Romeo Corpen 005
13 knots

Higgins

Course 185
speed 13 knots

Milius

Yukon

waiting
station

Elliot

Figure 9–1. USS *Higgins* (DDG-76) during underway replenishment.

4,500 yards. At that point I'm going to increase speed to eighteen knots and come right until the bow is pointing just astern of *Yukon*."

The captain interrupted. "Astern? How far aft of *Yukon* will the bow be pointed?"

"It'll stay pointed at a moving imaginary point about fifty yards aft of *Yukon*'s stern. My strict vow is, 'I shall not bring the bow across *Yukon*'s stern until I'm almost in waiting station and I slow.'" The captain nodded and Ed continued. "I figure you'll be sitting in your chair as we approach"—Ed nodded in the direction of the captain's chair on the starboard side of the pilot house—"and I want to keep you in that chair. So as long as I keep the bow pointed aft of the oiler you'll stay in your chair."

"Well, that's a real 'Destroyerman' approach, all right, and I hope you'll keep your vow of the bow and keep me in my chair." The captain walked toward his chair. "You're right, you know, I'm going to sit here and watch, quietly, but if our bow goes forward of *Yukon*'s stern, I'll be standing right on top of you, and not quietly. That's 'Destroyerman,' too." Smiling, he sat down.

"Ben, what's the range to *Yukon*?" Ed asked the JOOD standing at the pilot house radar repeater as Ed took a bearing with the pilot house alidade.

The response was "5,500 yards," and Ed added, "Bearing 220" and continued to take bearings.

"Ed," the captain asked, "at what point are you going to come to Romeo Corpen and slow to thirteen knots?"

"When range to *Yukon* is 650 yards, Captain. I figure making turns for eighteen knots we'll be making more like sixteen and a half knots good because using standard rudder will slow us that much, and with 40 yards per knot surge distance we'll cover about 140 yards as we slow to thirteen knots. I want to be about 500 yards astern of *Yukon*, so if I slow at 650 yards we'll surge to about 500 yards astern of *Yukon* when we're at thirteen knots." Ed stepped quickly to the starboard wing of the bridge to make sure all was clear, then returned to the pilot house alidade.

"Bearing 225," Ed said a few seconds later. "Range 5,000 yards," the JOOD said.

Ed kept his head down and eyed the alidade. "Bearing 230. Range?"

As Ben responded with "4,500 yards," Ed ordered, "Right standard rudder. All engines ahead full for eighteen knots."

"Right standard rudder, aye, Sir," from the helmsman.

"All engines ahead full for eighteen knots, aye, Sir," said the lee helmsman as *Higgins* turned to starboard with a noticeable increase in power and speed. Helmsman and lee helm reported rudder and engine response, and Ed said, "Very well," followed by "Belay your headings."

"Belay my headings, aye, Sir," responded the helmsman.

Ed nodded to the bos'n mate and said, "Set the refueling detail," and to the phone talker, "Sigs, Bridge, any Romeo flag on *Yukon?*" People were scurrying about with muffled voices during the relieving processes as word was passed and bos'n's pipe shrilled. "No Romeo yet on *Yukon,* but she's showing Prep at the dip to port for *Milius.* Fifteen minutes for *Milius.*" Ed gave "Bridge aye" and continued to watch the swing of the ship to starboard. The captain sat watching from his chair, binoculars in his hand.

"Ease your rudder to right 10 degrees," ordered Ed in a clear loud voice, pronouncing every word.

"Ease my rudder to right 10 degrees, aye, Sir." This was followed in a few seconds by "My rudder is right 10 degrees, Sir."

"Very well" was Ed's reply as the rate of swing to the right slowed, with the bow still not near the stern of the oiler.

"Ed, what are you going to use as 'waiting station'?" asked the captain.

"Three hundred to 500 yards astern of *Yukon,* off her starboard quarter, with 150 feet lateral separation, Sir."

"And how are you going to judge the lateral separation?"

"I'm going to use the 'WAR' method, Captain; wake, aspect, and radian. I'll be able to see *Yukon's* wake alongside us and I can use 'seaman's eye' to judge the distance. With some more 'seaman's eye' I can study the aspect of the oiler to see if it looks right. And there's the radian rule that tells us that at 500 yards range we should have *Yukon* 6 degrees offset to the left of Romeo Corpen to give us 150 feet separation. Three methods. I know you wouldn't want me to rely on only one method. But I kind of prefer the wake. We'll be able to use the laser range finder once we get alongside."

"I concur, Ed. Sounds like you've got it right. I prefer the wake, too, but never rely on only one source."

"Ease your rudder to right 5 degrees," Ed ordered. The ship's head was still swinging right, but much slower as it neared *Yukon's* stern. "Ben, give me a range to *Yukon* from that repeater every few hundred yards, and get a quartermaster on the stadimeter to do the same." To his phone talker, he said, "Combat, bridge, give me ranges to *Yukon* every two hundred yards. I want a number of sources."

Ben replied, "2,200 yards to *Yukon*," and the phone talker said, "Combat reports 2,300 yards to the oiler."

"Bridge aye," Ed acknowledged.

With the bow coming slowly toward *Yukon*'s stern, Ed ordered, "Ease your rudder to right 2 degrees."

"Ease my rudder to right 2 degrees, aye, Sir," followed by "My rudder is right 2 degrees, Sir."

Ed checked the rudder angle indicator and studied the ship's movement. The bow appeared to be headed for a point about fifty yards astern of the oiler, with ship, oiler, and imaginary point fixed, with no relative movement. Ed had brought the bow over to where he wanted and was now in a long continuous turn as the oiler proceeded on course 005, speed thirteen knots.

"Two thousand yards. Two thousand yards to *Yukon*." Then "1,950 yards." Reports came from three sources as Ed noticed a slight increase in the space between his ship's bow and *Yukon*'s stern. "Increase your rudder to right 4 degrees." The helmsman acknowledged the order and repeated accomplishment. "Very well." Ed sent the JOOD out on the port bridge wing to take bearings on the lifeguard destroyer *Elliott*.

"Left bearing drift on *Elliott*," Ben reported as Ed picked up the radio handset. "Echo Tango this is Hotel Sierra. I'm going to take waiting station five hundred yards astern of Yankee November, keeping ahead and clear of you. Over."

"This is Echo Tango. Go right ahead. Have a good drink. Over."

"This is Hotel November. Roger. Thank you. Out."

"Eighteen hundred yards . . . 1,600 yards . . . 1,400 yards." "Increase your rudder to right 7 degrees." "Twelve hundred yards." "Increase your rudder to right 10 degrees." Helm response. "Very well," Ed said as he saw that he needed more rudder to keep the bow pointed at that moving imaginary target. "One thousand yards." Ed directed his three range sources to notify him when the range to *Yukon* was 650 yards.

"Romeo at the dip to starboard on *Yukon*," sang out the signalman as Ed left his position in the pilot house, stepped out on the port wing, and stood behind the bridge wing pelorus. "Eight hundred yards."

"Put our Romeo at the dip to port," Ed ordered, and the signalman who had been peering from around the rear of the pilot house ran to his flag bags, shouting, "Dip Romeo to port." The ship was still swinging right, now with 10 degrees rudder, making turns for eighteen knots. The captain was watching carefully from his chair as his ship closed as if to pass astern of the oiler.

"Six hundred fifty yards," came reports from three sources. Ed immediately ordered, "Right standard rudder. All engines ahead standard. Indicate revolutions for thirteen knots." Helm and lee helm repeated Ed's order and reported accomplishment. Ed said, "Very well," and *Higgins* swung rapidly to starboard as she slowed. "Steady on course zero zero five," ordered Ed as he checked the rudder angle and speed indicators. "Steady on course zero zero five, aye, Sir," acknowledged the helmsman. Ed responded, then ordered, "Belay ranges except stadimeter." Ed knew that the radar ranges were not accurate this close, and the quartermaster with the antique but still useful distance-measuring device was standing next to him.

"Steady on course zero zero five," sang out the helmsman. "Very well," Ed responded, then said, "Range?" "550," came from the quartermaster. Ed leaned over the bridge wing rail, eyeing the oiler's wake as it streamed down the port side with blue water showing between the wakes. "Looks good. About 150 feet, I'd say," he said to all nearby.

"I concur, about 150 feet. What's our range, Ed?" The captain had walked out on the bridge wing as Ed came over to Romeo Corpen. "Four hundred seventy-five yards, Captain," responded the quartermaster after a look from Ed.

"How about aspect and radian rule?" asked the captain.

Ed's eyes remained fixed on *Yukon*. "Looks good to me, Sir," he said. "We're well up on her starboard quarter." He paused as he studied. "Yes, Sir, I'd say the aspect looks right for 150 foot lateral separation. And the radian rule, we're a little less than five hundred yards, so I'm dropping back." He ordered a reduction of 5 rpm in speed. "When we get to five hundred yards I'll check the bearing."

A signalman called out, "Prep hauled down on *Milius*. Completed fueling." They watched as with a swirl of white water the destroyer that had been fueling on the port side of *Yukon* increased speed smartly, moved ahead and gradually turned away.

Everyone's attention remained fixed on the oiler, and a minute later the quartermaster reported, "Five hundred yards." Ed ordered an additional 7 rpm with his head down on the alidade taking a bearing. "359 on the starboard side, that's 6 degrees left, or less, than Romeo Corpen." Ed looked at the unfolded paper diagram he had taken from his pocket. "Yes," he said. "Right on. Six degrees will give us 150 feet separation at 500 yards." He turned to the captain, smiling. "All three give us 150 feet separation, Captain."

The captain also smiled, still looking at the oiler. "Good. Either skill or luck, either one is good. Okay, are we ready to go alongside?"

"All stations report manned and ready for UNREP," the JOOD reported.

"Romeo is still at the dip," Ed said, pointing toward *Yukon.* "Belay that," he said. He watched as the red and yellow Romeo flag went close-up on *Yukon* and a signalman sang it out.

"Yes, Sir, we're ready, Captain, and *Yukon* is ready to receive us. Permission to go alongside?"

"Permission granted."

"All engines ahead full. Indicate revolutions for eighteen knots." Ed's order was acknowledged by the lee helm, who then reported the engines' response and the ship noticeably increased speed, closing on the oiler. The quartermaster with his stadimeter called out, "Four hundred fifty yards."

Ben moved closer to Ed and asked quietly, "How about our Romeo close-up? Aren't we commencing our approach?"

"Not yet. Waiting station is three hundred to five hundred yards astern of the oiler, and I'll two-block Romeo at three hundred yards. We'll be up to speed then and that'll make a shorter time from Romeo close-up to first line over. Looks better that way."

"Four hundred yards," the quartermaster called, then 350, then 300.

"Close-up Romeo," Ed called out as a signalman who had been holding a halyard repeated the order and the red and yellow flag jumped quickly from its "dipped" position to the yardarm. Ed leaned toward the pilot house door and announced, "We're commencing our approach on *Yukon.*"

"Fo'c'sle, bridge," he said. "Report when our bow crosses the oiler's stern."

"Two hundred yards . . . one hundred yards . . . fo'c'sle reports bow to stern."

Ed counted eight seconds then ordered, "All engines ahead standard, revolutions for thirteen knots." The helmsman acknowledged the order, reported accomplishment, and Ed answered, "Very well."

"Fo'c'sle, report lateral separation using laser range finder." Now they would learn if the three sources, wake, aspect, and radian rule, were accurate. The report, "160 feet," brought smiles on the bridge and the ship very gradually slowed as it moved forward parallel to and alongside *Yukon.*

"Why the eight-second delay?" Ben asked.

"Last week with the same five-knot speed differential we cut our speed just as our bow crossed the oiler's stern and we came up about eighty feet short. So this time I held on for eight seconds, and now we'll see if we match speeds at the right place." They continued moving forward alongside the oiler but at a considerably reduced relative speed.

"Come left, steer course zero zero four." Ed ordered a 1-degree course change to bring them closer. As soon as the helm reported steady, he went back to, "Come right, steer course zero zero five," and the fo'c'sle laser report was 150 feet. The ship matched *Yukon*'s speed near the forward fueling station, and Ed added two turns to move forward a few feet with the order "Indicate six zero revolutions." The lee helmsman responded, "Indicate six zero revolutions, aye, Sir," followed by, "Engines answer six zero revolutions for two revolutions over thirteen knots, Sir." "Very well," said Ed.

Yukon greeted them alongside on an announcing system and, after an appropriate warning, shot lines came across to start the rigging and refueling. Ed ordered, "Execute Romeo," and the flag jumped down. A few minutes later the rig and the phone and distance line were in place, and as hoses came over, he alerted his helmsman before the span wires were tensioned and watched as the ship carried 5 degrees right rudder to maintain heading. Hoses came across bringing fuel to the thirsty destroyer as Ed adjusted course with 1-degree changes to maintain separation and adjusted speed with 2 rpm changes to maintain fore and aft position. "One degree, two turns and three; a lot of patience." The captain chatted with *Yukon*'s master over the phone and distance line.

Half an hour later, fueling control station reported fifteen minutes to completion and Ed ordered, "Prep at the dip to port." When fueling had stopped and the last span wire was ready to be tripped, the prep pennant was quickly two-blocked.

"Fueling station requests permission to disengage final wire," came the report from the phone talker. Ed looked at the captain, who nodded as Ed responded with, "Permission granted to disengage," and permission was granted for fo'c'sle to take in the phone and distance line.

As the phone talker reported, "All lines clear port side," Ed ordered, "Sigs, execute Prep," and a signalman yanked a halyard taking the pennant from the yardarm.

"All engines ahead flank, revolutions for twenty-seven knots. Come right, steer course zero zero seven. Fantail, let me know when our stern is clear of *Yukon*'s bow. Bos'n, secure the refueling detail. Combat, recommend course to next assignment." Ed was still giving orders and receiving responses as he walked briskly from the port wing of the bridge, across the pilot house to the starboard wing. After ensuring that all was clear to starboard, he ordered the ship to 009, and when fantail reported the stern clear of *Yukon,* he ordered, "Right 10 degrees rudder." *Higgins* leaned slightly as she turned, increasing speed, carving an arc of white wake in the smooth blue sea.

The captain went to his seat and picked up a cup of coffee. "Nice job, Destroyerman." *Higgins* proceeded to her next assignment.

The ability to replenish naval vessels at sea is essential to forward deployment and to maintaining ships on station for long periods of time. It is said that when military amateurs get together they discuss strategy, but when military professionals gather, they discuss logistics. Underway replenishment, or UNREP, is key to naval logistics. The techniques of large-scale underway replenishment were developed by the U.S. Navy during World War II and have evolved steadily since that time.

There was some early experimentation in fueling bow to stern, both with both ships lying to, and with one ship towing the other. This did not prove as effective as alongside replenishment. UNREPs for fuel, stores, and ammunition are now almost always conducted with the ships steaming on parallel courses and linked up alongside. Allied navies now in many cases share the same procedures. Guidance is set forth in APP 4 Allied Maritime Message Formats, ATP 16 Replenishment at Sea, Fleet Replenishment Guide, NWP 3-04.1 Standard Helicopter Operating Procedures, NWP 4-01.4 Replenishment at Sea, NWP 14 (rev. D) Replenishment at Sea, and SORM (OPNAVINST 3120.32C).

Underway replenishment follows an established pattern. The officer in tactical command orders a course and speed (Romeo Corpen and speed) for the evolution. The delivering ship comes to the ordered course and speed, and signals on which side the receiving ship should approach. The entire operation is frequently conducted in radio silence, using flag hoists. The receiving ship moves into waiting station three hundred to five hundred yards behind the delivering ship, offset by the distance it will be while alongside, usually 120 to 200 feet, depending on the size of the receiving ship. When the delivering ship signals her readiness, the receiving ship increases speed and moves alongside. She then holds the position alongside while the transferring rigs are passed and tensioned, the transfers are completed, and the transferring rigs returned to the delivering ship. As the last line goes clear the receiving ship increases speed and clears out ahead, gradually increasing separation as it goes.

Flag hoists for underway replenishment are as follows:
Romeo at the dip:
On the delivering ship: Steady on Romeo course and speed, and am preparing to receive you alongside on side on which flag is flown.

Figure 9–2. Underway replenishment enables forward operations for extended periods. Here USS *Merrimack* (AO 179) refuels USS *San Jacinto* (CG 56) and USS *George Washington* (CVN 73). Because of the starboard location of the navigation bridge and the overhang of the angled deck to port, an aircraft carrier always takes replenishment station to port of the replenishment ship. *U.S. Navy photo*

On the receiving ship: I am ready to come alongside on side indicated.

Romeo close up:

On the delivering ship: I am prepared to receive you alongside.

On the receiving ship: I am commencing my approach to go alongside.

Romeo hauled down:

On both ships: Messenger in hand.

Prep (used only by the receiving ship):

At the dip: Fifteen-minute standby to breakaway.

Close up: I am disconnecting my last station.

Hauled down: All lines are clear.

Bravo: Handling hazardous material (fuel or explosives):

Closed up, both ships: Fuel or explosives are being transferred.

At the dip, both ships: Have temporarily stopped transfer.

Hauled down, both ships: Delivery of hazardous materials completed.

It is important to minimize the time spent in conducting an UNREP. Ships hooked up alongside each other for replenishment are more vulnerable both to accident and to enemy action. The chosen course for replenishment may take the ships away from where they need to be for operational reasons. Therefore we need to spend a minimum amount of time in going alongside, rigging for transfer, completing the transfer of fuel, stores and ammunition, and clearing from alongside.

Over a period of time, transfer rates have steadily increased. As the Navy entered the twenty-first century the underway replenishment system was able to transfer 4,000-pound loads by helicopter vertical replenishment (VERTREP) and 5,700 pounds in up to sea state 5 using a tensioned wire rope high-line transfer capability (conventional underway replenishment, or CONREP). Vertical replenishment is discussed in chapter twelve. Under special circumstances in lower sea states it was possible to transfer up to 10,000 pounds by CONREP. The maximum ship separation allowable was 180 feet. Plans are under way to develop a "Heavy Underway Replenishment System" capable of transferring at twice the rate of the current system, in higher sea states, and with a maximum separation between ships of up to three hundred feet. This system is scheduled to be ready for operational evaluation by fiscal year 2008. Even after this system becomes operational it does not necessarily mean wider separation between ships while replenishing.

Almost certainly the transfer rate will continue to be better when the ships are closer. The additional margin will permit ships to remain hooked up under circumstances that could require an emergency breakaway with current systems.

From the point of view of a shiphandler, underway replenishment is both challenging and rewarding. It can also be hazardous. The ships involved must steam one alongside the other in close proximity for an extended time. An error in judgment, a material failure, or a miscommunication can result in a collision. During an underway replenishment it is important to have our most skilled personnel on the bridge, in after steering, and in key engineering assignments. It is important to practice regularly in safe circumstances such evolutions as an emergency shift of steering control to after steering, so that all hands know what to do quickly and smoothly in the case of a real emergency. If time permits it is a good idea to conduct an emergency steering drill just before going alongside, with all of the same people involved who will be on watch while alongside. In the event of the loss of an engine the shiphandler needs to know what speed to order on the remaining shaft to maintain speed through the water, and how much rudder will be needed to compensate for the resulting unbalanced thrust.

Under most circumstances, during replenishment operations it is best to give the helmsman a course to steer, rather than rudder orders. The forces operating on the ship, which include weather, interaction between ships, and tension on the replenishment rig, can cause your ship to be carrying rudder, that is, to require an average rudder position of other than dead ahead to maintain course. An experienced helmsman will compensate for that.

For an underway replenishment the delivering ship maintains the signaled course and speed, with all of the maneuvering done by the receiving ship. Other considerations permitting, a course into the wind and sea is preferred. A following sea makes accurate steering more difficult, and a beam sea can cause excessive rolling. Operational requirements, however, often dictate something other than an optimum course, and it is the shiphandler's task to compensate. If on, or operating with, an aircraft carrier you can expect to do a lot of replenishments headed down wind as the carrier maneuvers to get sea room for flight operations. Normal UNREP speed is thirteen knots. This puts the speed above the range at which ships with controllable-pitch propellers adjust speed by varying the pitch angle of the propellers and into the range in which speed is controlled by changing shaft speed.

UNREP speeds up to twenty knots can be used, but as will be discussed subsequently, higher speeds make the evolution more difficult. Operational needs and safety parameters are often in conflict. It is the task of the officer in tactical command to weigh the relative importance of these and decide accordingly. It was not unusual on Yankee Station during the Vietnam War for a multiproduct replenishment ship to have an aircraft carrier alongside to port and a cruiser alongside to starboard, all steaming at twenty knots while the carrier operated aircraft, simultaneously transferring fuel, ammunition, and stores. In peacetime it would be difficult to justify the risks of such an evolution. In wartime, operational necessity sometimes overrules safety considerations.

The Approach

The approach is the most hazardous part of an UNREP. At least one commanding officer was in the habit prior to an approach of advising his conning officers that "too far is just too far. Too close is a collision." As with all shiphandling, the key is to plan properly the evolution in advance, measure and observe, and correct as necessary. In the theoretical ideal underway replenishment the receiving ship would start from precisely the waiting station assigned, would maintain the desired separation between ships during the approach, would slow to signaled speed at exactly the right time to settle in the correct position alongside, and would maintain precise separation and fore and aft position throughout the replenishment. In the real world we will have some adjustments to make along the way.

Of the evolutions involved in conventional underway replenishment, the approach places the greatest demand on the shiphandler. Sometimes the approach will begin from waiting station, which permits the shiphandler to set up the geometry of the approach and check to see if there is any difference between the signaled course and speed and the course and speed needed to maintain station on the replenishing ship.[1] Not infrequently, however, you will be called alongside without a pause in waiting station. The geometry is the same, but an approach without a pause in waiting station takes away your opportunity to observe the actual course and speed being made good by the delivering ship, and some extra time to set up your separation. Thus the direct approach is more demanding.

It is also possible to use a beam approach to go alongside the replenishing ship. In this case the ship being replenished starts from a position abeam of the replenishment ship and closes by steering a closing course 5 to 10 degrees off the signaled course, while maintaining bearing through speed adjustments. As

Figure 9–3. USS *Antietam* (CG 54) prepares to move alongside the Japanese supply ship JDS *Hamana* (AOE 424) to refuel. *U.S. Navy photo*

the range closes between ships, the approaching ship adjusts course in increments toward the base course. Most naval shiphandlers consider the beam approach to be somewhat more hazardous than the conventional approach from astern because of the momentum toward the delivering ship. In most circumstances the time saved is minor.

In the conventional approach from waiting station, the goal is to maintain throughout the approach the intended distance alongside. In most cases this will be 140 feet for destroyers, frigates, and cruisers and 150 to 200 feet for aircraft carriers and large amphibious ships. Once experience has been gained in making the approach, it can be useful to be somewhat closer during the first few minutes to make it easier to pass lines and hook up the rigs. It is a good idea to be a little wider before giving permission to tension the rigs, since this can set up another force tending to bring the ships together. Once alongside, a phone and distance line is passed which not only facilitates direct communication with the other ship's bridge, but also aids in maintaining precise distance (see fig. 9–4).

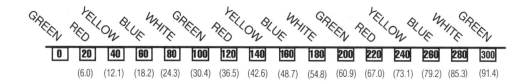

CONTROL SHIP   NOTE: FIGURES IN PARENTHESES ARE IN METERS   APPROACH SHIP

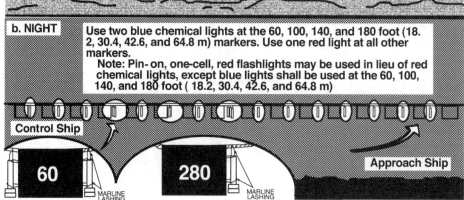

Figure 9–4. Phone and distance line markings.

The key to a good approach is continuous measurement of your position relative to the delivering ship. As with all shiphandling it is best to have multiple sources of information. There are three good means of doing this, identified by the acronym WAR. The letters in the acronym stand for wake, aspect, and radian.

An excellent way of checking on your distance during the approach is to observe the distance between the edge of the wake of the delivering ship and your own wake. With 5 feet of blue water between the wakes, your lateral separation is about 90 feet. With 40 feet of blue water, you are at about 125 feet. If the wakes of the two ships are touching, you are too close and it is time to alter your course outboard.

A second way of checking your distance is by observing the aspect of the delivering ship. While you are in waiting station you should be able to see quite a bit of the side of the delivering ship, not just the stern. As you move ahead during your approach, you should increasingly see more of the side, and less of the stern. Accurate judging of aspect requires a certain amount of experience, but is a valuable double check on the other ways of judging separation.

The third, and perhaps most precise, means of judging the separation is by making use of the radian rule. The radian rule takes advantage of the fact that for small angles the lateral offset is sixty times the degrees of offset.[2] Using the radian rule it is possible to construct a table. By entering the table with the range and bearing to the delivery ship you can determine what your separation will be.[3] The radian rule table for UNREP approach follows. (Enter the table with range in yards and angle in degrees to the near side of the delivering ship; this gives the separation in feet. The table may be extended as desired, using the radian rule.)

Angle in Degrees	1.5	2	3	4	5	6	7	8	9	10
Range in Yards				Lateral Separation in Feet						
1,000	75	100	150	200	250	300	350	400	450	500
900	68	90	135	180	225	270	315	360	405	450
800	60	80	120	160	200	240	280	320	360	400
700	53	70	105	140	175	210	245	280	315	350
600	45	60	90	120	150	180	210	240	270	300
500	38	50	75	100	125	150	175	200	225	250
400	30	40	60	80	100	120	140	160	180	200
300	23	30	45	60	75	90	105	120	135	150
200	15	20	30	40	50	60	70	80	90	100

For example, if at five hundred yards the replenishment ship as observed from our bridge bears 5 degrees to the left of base course, our offset is 500/60 × 5, or approximately forty-two yards. Since we measure separation in feet while alongside, our separation in this circumstance is 125 feet. In measuring bearing, make sure to take the bearing to the near side of the replenishment ship. Otherwise you will be closer than intended. Do not try to use radar bearings during the approach. Most radars will present the target as a large blob at the ranges involved here, resulting in imprecise bearings. Radar ranges can be useful, but remember that the range you get will be to the closest point of the target, and what you want to know is the distance to the alongside position. A laser rangefinder, or even a stadimeter will give you more precise ranges. It is useful to construct a table for your own ship, giving the desired bearing for each one hundred yards during the approach.

A good shiphandler must always be a pragmatist. It is not unknown for a replenishment ship (or your own ship: you did check, didn't you?) to have a gyro error, such that the true course being steered is different from that signaled. Signals are sometimes misread. At least one collision has been caused by a replenishment ship steering by magnetic compass rather than gyro. The point is that the conning officer cannot become complacent. If you are too close, and a course 5 degrees to the right of the signaled course is not opening the distance, come farther right. If the signaled replenishment speed is thirteen knots and you are still moving ahead with twelve knots rung up, drop to eleven knots. Believe what your observation tells you and correct accordingly.

Maintaining proper separation during your approach is only half of the evolution. The other part of the approach is controlling your speed to settle neatly into the desired position alongside. The approach is made at a speed from five to ten knots above the signaled speed. The larger your ship, the more momentum. This means that a large ship will normally use a smaller speed differential than a ship of lesser tonnage. For all ships it is a good idea to adopt a standard situation: always make the approach at the same speed increment above signaled speed. This way knowledge gained from each approach is cumulative and you will soon learn the exact point at which to slow for each of the types of replenishment vessels.

The relative distance your ship travels between the time you ring up the signaled speed and the ship actually slows to that speed is called the surge distance. No two ships will decelerate at exactly the same rate, so experience is needed to determine just when to begin slowing from approach speed. Typically, a destroyer or cruiser making an approach at eight knots above signaled speed should ring up base speed as the bow of the cruiser comes abeam the stern of the replenishment ship. An aircraft carrier will need to slow much earlier. It is useful during this phase of the approach to have someone singing out the ship's speed through the water from the pitometer log.

If it appears that you will overshoot or undershoot your desired position alongside, the first engine orders should be fairly bold ones. Once settled alongside, you will maintain station by ordering speed adjustments of two or three revolutions a minute, but not now. Your first speed adjustment can be two or three knots above or below base speed. It is particularly undesirable to overshoot by more than half of the ship's length, because in this position the ships are both being pulled together by the Venturi effect and rotated toward each other. If necessary, before you get in this position, it is safe to adjust your speed by as much as five knots above or below base speed, but do not wait to see it actually taking effect before resuming something closer to base speed. Otherwise you can get into oscillation.

A useful technique is "bracket and halving." To do this you use a big enough speed adjustment to start moving in the desired direction, then follow it up by an engine order of half the magnitude in the other direction, progressively reducing the size of the engine order increments until you are settled alongside. Once settled alongside, there is little interaction between speed and the course being steered. Some commanding officers take advantage of this for training purposes by "splitting the conn," with one officer controlling distance between ships by orders to the helm and another controlling fore and aft position by orders to the engines. Most do not like the ambiguity as to who really has control of the ship.

Alongside

Ships steaming in close proximity, as during an underway replenishment, interact with each other in a significant way. Bernoulli's principle (see chapter 2) tells us that low velocity of a fluid is associated with high pressure, and that high velocity is associated with reduced pressure. As a ship moves it displaces water to either side, with the water rejoining at the stern. It is, therefore, evident that the water passing down the side of the ship must take a longer path, and move at a higher velocity. The result is high water pressure at the bow and reduced water pressure along the sides of the ship. When ships are steaming in close proximity, the effects can be complex (see figs. 9–5, 9–6, and 9–7).

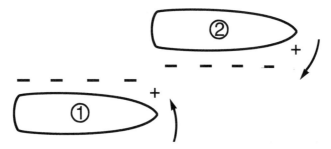

Figure 9–5. During the approach, as ship 1's bow comes abeam of ship 2's stern, her bow is drawn toward the low-pressure area at ship 2's stern. At the same time, the high-pressure area at ship 1's bow repels the stern of ship 2. The resultant force tends to make the ships yaw toward each other.

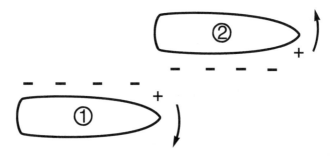

Figure 9–6a. As the ships overlap, the high pressure at ship 1's bow tends to cause the ships to yaw away from each other, though in practice this effect is not reliable.

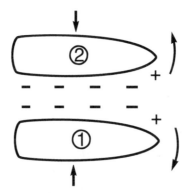

Figure 9–6b. As the two ships come abeam, the bow pressure area between the two hulls exerts a strong attractive force. The combination of high pressure forward and low pressure aft induces a strong outward yaw force. This combination usually results in the ship maintaining station having to use courses away from the ship on base course while the helmsmen on both ships carry rudder toward the ship alongside to maintain the ordered courses.

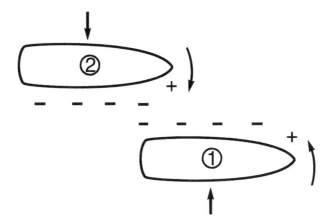

Figure 9–7. If ship 1 overshoots her position alongside ship 2, the ships are still attracted but are now trying to yaw toward each other as well. This is a dangerous position because of the combination of the ships being both pulled and rotated toward each other.

If two ships are moving alongside each other there will be a region of low pressure between the ships. The closer the ships are together, and the higher their speed, the lower the pressure of this Venturi effect will be. The effect increases exponentially with higher speed and reduced separation. This creates an uncomfortable situation, in that the closer two ships alongside each other come together, the more difficult it becomes to open back out. Most shiphandlers consider 60 feet to be the point of hazard. With speed held constant, the forces pulling the two ships together at 60 feet are four times what they are at 120 feet. Similarly, at twenty knots, the forces are four times as much as at ten knots.

Another force that will affect your ship while alongside is the tension on the underway replenishment rigs. Stations will request permission before tensioning the rigs. NWP 14 requires that the distance between ships be no closer than 140 feet before tensioning. It is probable that after tensioning that the amount of rudder the helmsman has to carry to maintain his ordered course will change, and he should be alerted ahead of time. Rig tension may also apply a rotating force, particularly if only the after rig is tensioned. This is likely to change the ordered course needed to maintain distance.

Wind can also be a complicating factor while alongside, particularly if there is a strong wind from the beam. The ship which is to windward will experience the full force of the wind, and thus a set to leeward. The downwind ship is at least partially shielded by being in the lee of the other vessel, and thus experiences less set. A beam wind thus tends to force the two ships

Figure 9–8. Weather can make underway replenishment more difficult. Note Romeo close up on the destroyer as she makes her approach alongside. *U.S. Navy photo*

toward each other, no matter which ship is to windward, and requires extra attention to maintain proper separation while alongside.

Changing Course and Speed

Not infrequently it will be necessary to change base course and/or speed while alongside. This is relatively simple to do. The key is good communication between the two ships. The maneuver is controlled by the delivering ship, which announces course changes in increments, usually 5 or 10 degrees at a time. So long as a steady distance is being maintained between ships, it is not necessary to steady up on a new course before announcing the next increment of turn. If making large turns, however, it is a good idea to steady up every 45 degrees. Speed changes are even easier, and changes of one knot at a time can easily be followed by the ship alongside.

Emergencies

Potential emergencies while alongside include loss of steering control, loss of propulsion, and failure to maintain proper separation between ships. All of these situations can call for an emergency breakaway from alongside. The goal in an emergency breakaway is to disengage quickly without harm to personnel, the ships, or the rigs. Given a choice, damage to the rig is always less serious than a collision, although a tensioned rig should not be disconnected before tension is released. If unable to clear from alongside, the goal is to minimize relative motion between the ships. If only two ships are involved it may be possible to maneuver to avoid damage.

In the case of loss of propulsion power by either ship, if steering control is maintained the ship retaining power can probably adjust speed to maintain relative position until the emergency breakaway is complete. If power is lost on one shaft of a multishaft ship it should be possible to maintain signaled speed using the remaining shaft(s). The helmsman will have to carry rudder to compensate for the offset propulsion. Every ship should have a table of one shaft speeds posted on the bridge. When circumstances permit it is a good idea to give the ship's master helmsmen an opportunity to practice steering while dragging a shaft.

If steering control is lost, the ship with maneuverability may still be able to maintain separation. In this kind of emergency it is vital to maintain good communications between the ships. The ship with the casualty should announce speed through the water and ship's head at short intervals, to let the other ship maneuver to maintain position.

In 1995, there was a collision between a CVN and an AOE during an UNREP. The AOE had a steering problem, and the CVN adjusted course to maintain the distance between the vessels. When the AOE regained steering control, it returned to the signaled replenishment course, but without telling the CVN. The resulting collision caused more than $5 million in damage to the two ships. The collision probably could have been prevented by better communication between the ships to let each know what the other was doing. When coping with an emergency like a loss of steering control on board your own ship it is hard to remember to keep the other guy informed, but it can make the difference between a safe recovery and a collision.

Loss of steering control can leave little time for recovery. Experiments to determine how much time was available under various circumstances were conducted on the USS *Vincennes* (CG 49) and reported in the Naval Safety Center's publication *Fathom*. The experiment used towed barrels as the "delivery vessel." The results reported were that a rudder jammed toward the delivery vessel at 10 degrees resulted in a collision in about forty seconds; a rudder jammed at 30 degrees caused a collision in twenty-eight seconds. Circumstances of the test necessarily did not include the Venturi effect between ships, nor the ability of a real delivering ship to maneuver to try to maintain separation. The test did show that the time available to react is short, and that actions to regain steering control need to be rapid and well rehearsed.

As part of the same series of experiments *Vincennes* tested the often discussed tactic of twisting the ship's engines to offset a jammed rudder. Their finding was that an ahead flank/back two-thirds twist could not offset a 10-degree jammed rudder. It did, however, slow the rate of swing, giving more time to correct the steering problem. The lessons learned as reported by the commanding officer of *Vincennes* were, "Every second counts. First, you have to be proactive. Second, helmsmen and safety observers instinctively must execute planned responses. Last, you have to practice."[4]

Another emergency that can occur while alongside is to get too close. The exponentially increasing Venturi force tending to bring the ships together means that once you get too close, prompt and decisive action is needed to regain a safe distance. Because all shiphandlers internalize the lesson that the stern swings further than the bow, there is sometimes unwarranted reluctance to use enough course change to salvage the situation if the ships are too close. Once the distance between ships is less than sixty or eighty feet, the Venturi effect is increasing sharply and you need to change course enough to open the range promptly. As a rule of thumb, the stern of your ship will swing no more than ten feet toward the replenishment ship for every degree your ship's head is pointed away. Therefore, a 5-degree course change will swing the stern no more than about fifty feet, and this should be enough to start opening the distance between ships. Unless you know how much rudder the helmsman has been carrying to maintain position alongside, this maneuver is best accomplished by an order to come smartly the 5 degrees to the new course. This will probably be enough to salvage the situation. It is now necessary to exercise caution that you do not open too far. As soon as you can see the distance between ships start to open, start coming back to the base course a degree or two at a time.

Leaving Alongside

Fifteen minutes before completing replenishment, PREP is hoisted at the dip on the replenishing ship to inform the next ship scheduled alongside. PREP is closed up as the rig is disengaged, then hauled down as the last line goes clear. The clearing ship increases speed to clear from alongside, changing course by small increments away from the replenishment ship. Larger rudder angles should wait until you are well clear. Know well ahead of time where you are headed next. Move to the wing of the bridge toward which you are turning. As always, planning is the key to competent shiphandling. If it is necessary to cross over the bow of the replenishment ship make sure you are well clear ahead. Most experienced shiphandlers will want at least two thousand yards before passing ahead (three thousand yards if ahead of an aircraft carrier), and even then it is preferable to cross at a shallow angle until fully across the replenishment ship's bow.

10

SHIPHANDLING IN
EMERGENCIES

◄◦►

Dear Dad,

Remember your "man overboard" off *Buchanan,* Vietnam time, Gulf of Tonkin? You've told me that sea story lots of times and we've discussed the procedures. Well, now I can match that because yesterday I had the conn when a man went over the side and the captain let me maneuver the ship for the pickup. Maybe we're the first father and son to conn ships for man overboard recoveries.

We had just completed refueling, had cleared the oiler and the formation, and our UNREP detail was being secured. I was OOD in the regular 1200–1600 underway watch, so I relieved Matt Moran and took the conn. Captain Richardson had gone below and we were on course 212, making twenty-two knots en route to rejoin the *Stennis* Battle Group.

One of the deck force seamen was replacing a section of deck-edge lifeline that had been removed for refueling. The ship rolled and he fell overboard! Fortunately, he still had on his life jacket and MOBI and others of the deck force saw him go. I was out on the port bridge wing and heard them, the after lookout and the phone talker all yelling, "Man overboard, port side."

I immediately ordered, "Left full rudder." It was an automatic response, a reaction, just like we had practiced, and as I looked aft I could see him in the water and could see the stern start to move away from him. At twenty-two knots the ship was past him in about five seconds and he was clear of the stern. There were two other life jackets that had been thrown near him and a smoke flare that took a few seconds to light off.

Just as we had practiced, the JOOD Cliff Deets ordered the word passed, sounded six short blasts, and had the signalmen break Oscar. All of a sudden, Captain Richardson was at my side just as the phone talker identified the man overboard as Seaman Eric Plocica. The captain told Cliff to notify the XO that we didn't need a muster, we had identified the man.

We could see Plocica in the water on the port quarter facing us, making no movement, and as the ship curved away from him, the phone talker spoke out, "Combat reports true wind from 307, sixteen knots." I asked Combat to verify true wind and told them we had the man in sight.

"Looks like we can make a shipboard pickup. What do you think?" the captain asked.

"Yes, Sir," I agreed. "Shipboard pickup." And as the ship swung past 30 degrees left of our original course, the helmsman called out, "Passing 180."

The captain asked what kind of turn I intended. "I was going to do a Williamson Turn but the man is in sight and I think a Racetrack will be faster," I said. "Also, an Anderson might be even faster but could put us alongside him heading nearly into the wind, but the Racetrack should put him on our port side with the wind on our starboard beam."

When the captain nodded and said, "Sounds good. Go ahead with it," I realized that he was going to let me conn the ship and make the pickup, that is, until I did something wrong. He's a real fine shiphandler, you know, and I had assumed that he would take the conn for an actual man overboard, but just like we had practiced, he let the OOD make the recovery. And the recovery went just as we had practiced.

"Passing 160," the helmsman called, and I kept the rudder at left full as we passed the point where a Williamson Turn would have meant a shift to right full rudder. "Passing 140 . . . passing 130." We continued the left turn toward 032, the reciprocal of the original course. Plocica's relative position had moved steadily forward from deep on the port quarter to almost on the port beam, range six hundred yards. We could see only his head above water with no movement apparent, and he seemed to be facing southwest, the same direction as when we had first seen him.

Captain Richardson didn't lower his binoculars as he spoke: "He may be in shock or maybe he hit his head as he went over. Let's get the leading petty officer from where he went over, or the best witness so we'll know what to expect. We'd better be prepared for a full pickup because I don't think he'll be able to help himself." Cliff went to the phone to carry out the captain's orders.

As ship's head reached 047, 15 degrees before the heading I wanted, I ordered the rudder amidships and the ship's turn slowed. Nearing 035, I ordered, "Steady on course 032," and we paralleled the original heading on the backstretch of the racetrack. Still making turns for twenty-two knots, full rudder had slowed us to about nineteen knots, but now, steadied up, we increased to the full twenty-two knots.

With Plocica bearing 302, abeam to port, I counted ten seconds, ordered, "Left full rudder," and the ship started a sharp turn back to our original course. I wanted to come out of this turn in the wake we had laid in the water when Plocica went overboard. I could clearly see the smoke marker and barely see nearby debris, but two signalmen and the Captain had their binoculars fixed on the immobile guy in the water. "Good," the captain said when I put over the rudder.

"Passing 020 . . . passing 010." I was fixed to the pelorus watching ship's heading and the wind, and the helmsman called out every 10 degrees as the ship came around.

The captain and Cliff had a discussion with BM2 McKearney, and the captain seemed to confirm that Plocica had gone overboard "clean" but was probably in shock. We couldn't expect that he could grab a life ring or put on a lifting collar, and the fo'c'sle detail was rigged for pick up on the port bow with swimmers ready.

We were passing 290 when I started to pay more attention to the ranges that were coming from the phone talker. "One thousand yards. Nine hundred fifty yards. Nine hundred yards."

"Passing 280. Passing 270. Passing 260."

"Eight hundred fifty yards." "Passing 250."

"Eight hundred yards." "Passing 240."

"Seven hundred fifty yards." "Passing 230."

When I heard, "Seven hundred yards" and "Passing 220," I ordered, "Rudder amidships. All engines stop." I noticed the captain glance at me, and as soon as the lee helmsman acknowledged the engine response, I ordered, "All engines back two-thirds." The captain returned his eyes to Plocica, almost directly ahead, fine on the port bow, and I saw that with no rudder and both engines backing the ship would steady on about 210. Just what I wanted.

The ship was slowing with a slight shudder as we approached Plocica with the wind on our starboard beam. He was on our port bow still facing south west, that is, with his back to us, the same direction as when he went over, and he was making no movement.

"Three hundred yards."

"Two hundred yards." The ship was making about four knots still slowing. I secured radar ranges and asked the fo'c'sle for ranges.

"One hundred yards on the port bow." I could see the chief bos'n mate on the fo'c'sle at the jackstaff bracket pointing, but Plocica was hidden to us on the bridge wing by the flare of our bow. The ship was making less than a knot, and I stopped both engines as we crept closer.

Fo'c'sle reported no response as life rings were thrown near Plocica and I could hear the deck force seamen, his buddies, yelling at him. No movement. No response. The ship settled into position with Plocica less than fifty yards off the port side of the bow and the wind very slowly moving the ship toward him.

Fo'c'sle asked for and received permission to put two swimmers in the water, and in short order Plocica was lifted to the deck in a horse-collar harness, wrapped in blankets and hustled to sickbay. Our swimmers even recovered the additional life jackets that had been thrown over, and soon we were under way as before, en route to rejoin the *Stennis* Battle Group.

Chief Corpsman Elder reported from sickbay that Plocica was in good condition. He had mild hypothermia and had gone through some shock but he would be fine. Just before Captain Richardson went below to sickbay, he said that I had done some fine shiphandling and that the entire team had done a great job. Of course, with appropriate modesty, I said, "Just like we trained, Captain."

(Oh yeah, Dad, MOBI is a new device we have. It's a man overboard indicator system. If a man falls overboard, it sends out a signal so we can locate him.)

Going to sea remains an inherently hazardous business. A large part of the shiphandler's art is knowing how to handle best the variety of emergencies that can occur at sea. Some of the emergencies touched upon in this chapter include man overboard, towing, rescue at sea, and heavy weather.

Aircraft and Man Overboard Recovery

Proper response to a plane or a person in the water depends on circumstances. Is the report immediate or is it time late? Do we know what side the person went overboard? Is the person in sight? Taking weather and facilities available into account, do we want to recover by helo, by boat, or by shipboard recovery? Are there any navigational hazards to take into account? If other ships are in company, ATP-1, volume 1, chapter 5 provides maneuvering instructions.

If the report is immediate, and the side known, the first action in all of the recovery methods is to use full rudder to swing the stern away from the person in the water. Some shiphandlers advocate stopping the shaft on the side the person went over. With an instant report and instantaneous action this can be useful, but in most cases it is too late. Note, however, that it is not necessary to actually stop the shaft: if it slows even a limited amount suction comes off the screw. Ring up maneuvering combination (usually 999) on the engine order telegraph in anticipation of using your engines in the maneuvers to come. Simultaneously, life rings, life jackets, or just about anything else that floats should be thrown in the direction of the person. This serves the dual purpose of helping the person to stay afloat and marking the location visually. Smoke floats should also be deployed to mark the spot. At night a light is crucial. It is surprisingly easy to lose sight of a person in the water, even in good visibility and relatively calm seas. CIC should be directed to mark the location, to be able to give continuous ranges and bearings to the spot. Ship's location by GPS should be recorded immediately. As soon as a person in the water is seen, it is vital to assign one or more persons with no responsibility other than keeping the person in the water in sight, using binoculars if necessary. It is helpful to have this individual point continuously to the person in the water to provide a visual reference for the conning officer.

If a plane goes into the water ahead, as in the case of a cold catapult shot from an aircraft carrier, the most immediate concern is to not run over the crash site. The first action is to use full rudder to swing the bow away from the crash site, then a shift of rudder to swing the stern away as the crash site comes abreast the bow. Depending on how far the crash site is ahead of the ship, it may be possible to make a straight in approach, backing to stop abreast the plane and crew, keeping them to leeward.

Maneuvering for Recovery

The classic man overboard recovery maneuver is the Williamson turn, developed during World War II by Cdr. John A. Williamson, USNR. The goal is to reverse course in a way that heads a vessel back along its original track. When properly executed, the maneuver will place the ship in her own wake on a reciprocal course. The Williamson turn can be useful for recovery of a person whose position is known. It is almost imperative if the person's position is not known.

The classic Williamson turn starts with full rudder to the side on which the individual went over, if known. Continue with full rudder until the ship's heading is 60 degrees past the original heading, then shift to opposite full rudder. The momentum of the turn will normally carry the ship's heading to about 90 degrees from the initial heading, at which point the bow starts to swing in the opposite direction. Continue the turn with full rudder to steady up on the reciprocal of the initial heading. At this point the ship should be headed down the reciprocal of the original track about one turning diameter away from the point at which the turn started (see fig. 10–1).

Not all ships will respond identically to the Williamson turn, and it is important to calibrate your own ship to know at exactly what point to reverse rudder. A typical point to shift rudder for a destroyer is 60 degrees from the original course, yet it has been reported that with a very large crude carrier (supertanker) that the Williamson turn required that the rudder be shifted when the ship was only 35 degrees from the original course.[1] You should take the first opportunity in seas calm enough for the wake to persist as a visual reference to calibrate your Williamson turn. At normal transit speed, try the maneuver first shifting rudder at 60 degrees. If you turn inside your wake, try the maneuver again shifting rudder at 50 degrees. If outside, try shifting at 70 degrees. Once calibrated, post the data on the bridge. Note that engine speed is not changed

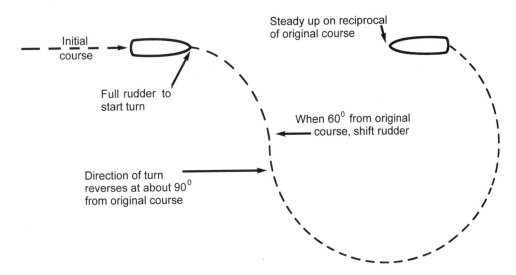

Figure 10–1. Williamson turn.

during the turn. The ship will lose speed during the turn, and generally the larger the ship the more speed will be lost.

If the person is in sight under good conditions and you are on an agile ship, it may be faster to use a continuous full rudder turn, sometimes called an Anderson turn. In this case, full rudder is used until the bow is pointed at the person in the water. Engines can be used to tighten the turn (see fig. 10–2). Some shiphandlers prefer to ring up full speed during the first part of this turn, but in any event it is important to get way off the ship early enough to avoid the risk of overrunning the person in the water. Backing power varies enormously between ship types, so it is again important to calibrate your own ship. As a rule of thumb, for a destroyer making sixteen knots at the start of the turn, try a back two-thirds bell when five hundred yards from the person in the water. Until you have this maneuver accurately calibrated it is better to take off speed a little early than to risk overrunning the person in the water. The back bell can be adjusted either up or down to bring the ship to a stop at the chosen pickup point. This will normally be with the person in the water a few yards to leeward of the ship's bow. The ship will drift downwind at a greater rate than the person in the water. Care is needed to keep the person

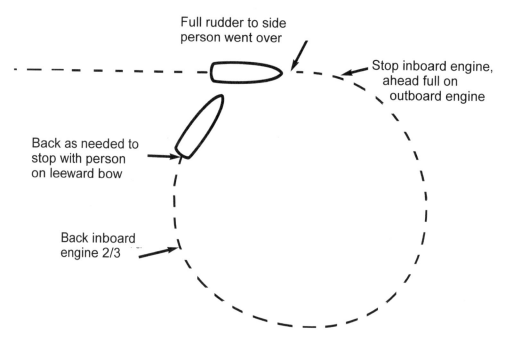

Figure 10–2. Anderson turn.

being recovered forward of the ship's intakes and screws. If it is necessary to put a swimmer in the water, ensure they have life jackets and a tending line.

Another alternative approach when the location of the person in the water is known is the racetrack turn (see fig. 10–3). As with the other methods, it begins with full rudder to swing the stern away from the man overboard. The turn is continued through 180 degrees to steady up on the reciprocal of the initial course. This course is maintained until the person in the water passes abeam. Another 180 degree full rudder turn should then put the ship into its own earlier wake and headed directly toward the person in the water. Although slower than the Anderson turn, the racetrack turn shares with the Williamson turn the advantage that it will bring the ship back to its initial point even if sight of the person is lost. It is also preferred when a sonar tail is streamed, since it is less likely to cause fouling than the other methods.

A fourth man overboard recovery maneuver is Y-backing. This is used principally by ships whose low height of eye limits their range of visibility. The Y-backing maneuver begins as do the other recovery methods with full rudder to swing the stern away from the person in the water. When well clear all engines are backed full. As the ship loses steerageway, shift the rudder. Back until the bow has swung enough toward the person in the water to complete the turn toward them with an ahead bell (see fig. 10–4). The rest of the

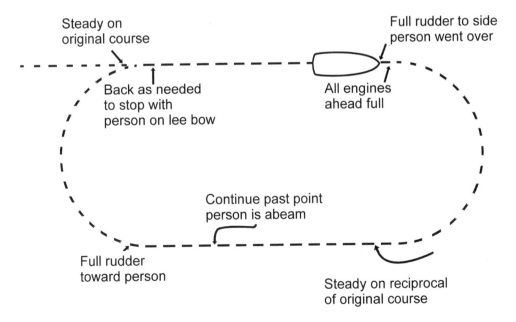

Figure 10–3. Racetrack turn.

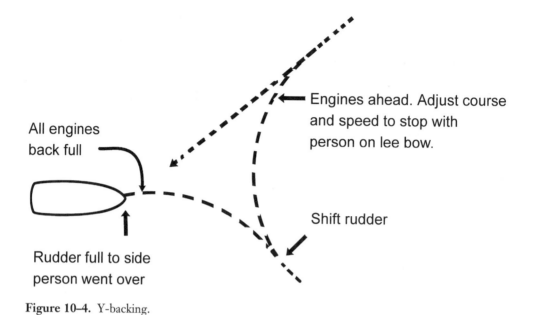

Figure 10–4. Y-backing.

pickup is completed as with the other maneuvers. The Y-backing technique is not good for ships with considerable sail area because of their tendency to back into the wind, making control more difficult.

There is no single best method for man overboard recovery. It is well to practice each of the alternatives to get a feel for what works best in a variety of circumstances. It is also an excellent means for student shiphandlers to gain a feel for the way the ship responds to engine and rudder orders, without the inherent risks of shiphandling in close proximity to other ships or navigation hazards. In a real man overboard situation, if in doubt use the Williamson turn.

Towing

Unlike a tugboat, which can spend most of its working life towing, for most surface ships it is an infrequent evolution. We will not cover routine tug towing here, but limit the discussion to the task of taking a disabled vessel under tow at sea. The need to tow is often the result of some emergency: accident, weather, battle damage, or mechanical failure. This adds a degree of difficulty. We will discuss the three stages in taking a disabled vessel in tow: passing a line, rigging the tow, and towing.

The first step is getting a messenger line from the towing vessel to the vessel to be towed. This requires placing the stern of the towing vessel at a distance of perhaps one hundred feet from the forecastle of the vessel to be towed. A caution: in exercise conditions, if the ship to be towed has variable-pitch propellers, its shafts may still be turning, and since zero pitch is rarely exactly zero, it is important to know if it has way on before getting close. If the wind is strong, and the ship to be towed is lying at an angle to the wind, it may also be making some way.

One classic shiphandling text recommends an approach similar to that used for an underway replenishment, with the towing vessel passing about one hundred feet to windward on a nearly parallel course (see fig. 10–5).[2] This can easily develop into a hazardous situation, however, particularly if

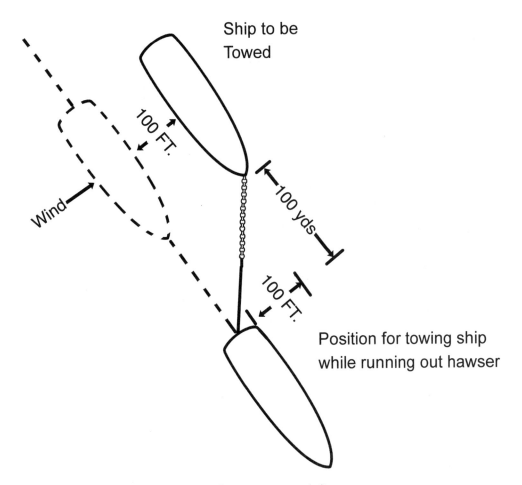

Figure 10–5. Towing: parallel approach (not recommended).

there is significant wind. Because the approach must be made slowly in order to stop with our stern about one hundred yards ahead of the vessel to be towed, a beam wind has a proportionately much greater effect than during an underway replenishment. In addition, as the vessel to be towed comes into the lee of the towing vessel it is less affected by the wind. The consequence can be that the towing vessel is set down on the vessel to be towed.

A preferable technique is to come in wider, but at a sharper angle, and aim your own ship's bow at a point about one hundred yards ahead of the disabled vessel. The quarter of your ship should pass the other's bow at about one hundred feet. Pass the heaving line at this time. As your stern comes abreast of the other ship, use rudder to swing your stern to control distance to their bow. Back to come to a stop with a separation of about one hundred yards between your stern and the other ship's bow (see fig. 10–6). Once the messenger is passed, it is important to keep a stable distance between your stern and the other's bow. Someone should be assigned to provide the bridge with continuous ranges using a laser range finder or a stadimeter. Caution needs to be exercised to keep your screws clear as the messenger is passed and

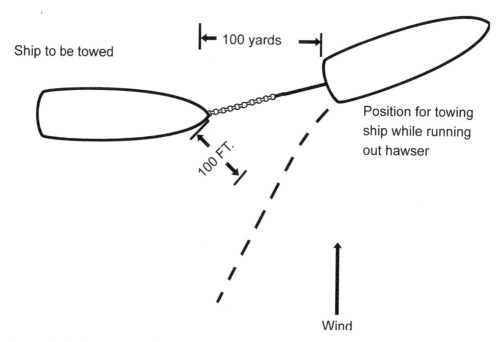

Figure 10–6. Towing: angled approach (recommended).

the tow is rigged. Make sure that excessive slack is not developed at any time.

As a general rule the longer and heavier the tow line the easier it will be to tow. The catenary provided by the weight of the tow line (usually provided by anchor chain from the disabled vessel) serves as a shock absorber. The larger the vessel to be towed, the more weight is needed in the catenary. A means of rapidly casting off the tow needs to be included, and a continuous watch stationed competent to do that. As weight comes into the towing rig it will tend to pull the ships together, so small ahead bells will be needed to maintain distance.

Once the rigging is complete, you are ready to start towing. The guideline here is gently. Sudden jerks or strains are to be avoided; they can easily part the rig. Under no circumstances should the catenary come completely out of the water. Navigational considerations permitting, start the tow headed in the direction the ship to be towed is already pointed. After gently taking all slack out of the towline, increase speed by a few turns at a time. You can then begin to work around slowly toward the heading desired. If power is available to the rudder of the ship being towed, they should steer to remain in the wake of the towing vessel. If power is not available the rudder needs to be placed amidships by whatever means is available. The length of the tow may need to be adjusted so that both vessels are riding the same to the waves.

Rescue at Sea

There are too many variables for it to be practical to cover all of the circumstances for rescue at sea. The most likely, and most difficult, are rescues that take place in heavy weather. For a person or persons in the water the considerations are similar to those for a man overboard. In heavy weather a major concern is the possibility of your ship injuring the person in the water as they are brought alongside. Even with the ship providing a lee, in heavy weather her motion can be violent enough to inflict serious injury. One technique that has worked well is to launch an inflatable life raft, which will ride much more safely alongside than will a rigid hull boat or a person in the water. If you need to put a swimmer in the water (always tethered and with a life jacket) to aid in the rescue, he can work more effectively from on board the life raft.

There are special hazards when assisting a ship in distress. It goes without saying that if the cause of the distress is grounding, then the danger of grounding exists for you as well. If the distress is caused by fire or by loss of propulsion you are going to have to work close to the distressed vessel, creating the risk of collision. We are used to going alongside other vessels on parallel courses during underway replenishment and when mooring inside a harbor. This is usually not the best approach in the open ocean. Ships in proximity parallel to each other tend to come together, either because the upwind ship is set down upon the leeward ship, or because they lie differently to the wind and rotation forces them out of parallel. It can be hard to extricate yourself from either of these positions. Circumstances permitting, it is much better to point your bow directly at the side of the distressed vessel. This allows precise control of the distance between ships by use of your engines (see fig. 10–7). In most cases it is best to approach from the leeward side, to take advantage of the lee created by the distressed vessel. As it is set down toward you it is easy to back gently to maintain the desired distance. If

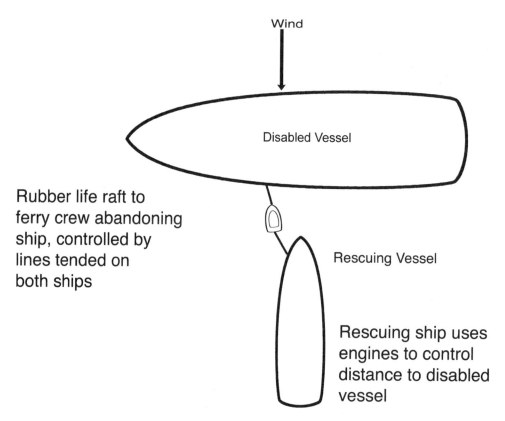

Figure 10–7. Maintaining position on a disabled vessel.

persons from the distressed ship are to be rescued, the preferred means is by inflated life raft under positive control by tethers to both ships. It is also possible to have the tether line passed through a block on the disabled vessel to permit the rescuing vessel to tend both the inhaul and the outhaul.

Do not, unless there is no alternative, go alongside another ship in an open seaway. Even in what may appear to be calm water there will be enough motion between ships to cause damage. If you have no alternative, use the biggest fenders you have.

Heavy Weather

There are too many variables for there to be a single set of rules for handling a ship in heavy weather, but there are some general principles that apply. The first rule is to avoid heavy weather. The steady improvement in weather reporting and forecasting makes this easier than it once was. Yet for many reasons we cannot always avoid storms. If a well-protected harbor is available, this may be the place to ride out a storm, although normally safe harbors can

Figure 10–8. USS *Paul F. Foster* (DD 964) turns away after an attempt at heavy-weather underway replenishment. Later on the same day, the replenishment was completed successfully. *U.S. Navy photo*

become dangerous with the winds and high tides that can accompany storms. History is the best guide to the harbor's safety in a storm. For most naval vessels the preferred choice is to leave harbor well in advance of a tropical storm to evade the storm if possible, and to ride it out at sea if necessary.

The second rule is to get sea room. The hazards of a lee shore are no longer what they were in the days of sail, but sea room still provides more choices and more safety. With sea room you have more freedom of choice in adopting a course to avoid or, if necessary, to ride out the storm. Even with vastly improved storm reporting, it remains prudent to make and compare your own observations. In the Northern Hemisphere, Ballot's law tells us that if you face directly into the wind that the center of the storm is about 115 degrees to your right. A constant wind direction with increasing velocity means that you are in the direct path of the storm. Wind that veers (shifts clockwise) means that you are in the "dangerous semicircle." A wind that backs (shifts counterclockwise) means that you are in the so-called safe semicircle. In the dangerous semicircle the winds are higher because the velocity of the motion of the storm is added to the wind, while in the safe semicircle the motion of the storm subtracts from the wind's velocity.

The rule of thumb for a vessel finding itself ahead of an approaching storm in the safe semicircle is to place the wind on the starboard quarter and proceed at the best speed at which the ship rides well. A ship in the dangerous semicircle lacks ideal courses of action. If far enough ahead of the storm, a fast ship can choose to outrun the storm to the point that she can safely move off the track. This carries the hazard that if the weather increases you may be unable to maintain the requisite speed of advance. For a ship in the dangerous semicircle that is slower or closer to the storm, the best tactic is choose a course as near as possible to a right angle to the storm's path to maximize the distance to the eye's closest point of approach. These rules are, of course, for ships in the Northern Hemisphere, and will be reversed for ships in the Southern Hemisphere.

Beyond the need to ballast in the interest of stability, there is no unanimity as to the preferred tactic if your efforts to avoid are unsuccessful and you are faced with the need to ride out a large storm. There are advocates for at least three alternatives. One is to steam into the wind at the minimum speed necessary to hold the ship's head. There are two variations of this tactic. One is to place the wind and seas about thirty degrees off the bow, on the argument that this reduces the pounding of heading directly into the sea. Alternatively, some advocate steaming directly into the sea, arguing that having the wind and sea on the bow requires a higher speed to hold the ship's head, with a resultant increase in pounding. In all likelihood, neither alternative is always to be preferred, and you should choose whichever seems to be working best in the circumstances. The second tactic is to run before the sea. There is no doubt that a following sea is more comfortable; the hazard is that the ship can yaw dangerously, or can be pooped by a following sea. This alternative is best for a ship that is longitudinally stable, and lacks any tendency to yaw or broach to. The third tactic with a substantial number of adherents is to lie to and let the ship find its own relation to wind and wave. The desirability of this course of action would again seem to depend on individual ship characteristics. It would not appear to be a good tactic for a ship that wants to lie beam to the wind, because of the possibility of heavy rolls when in the trough of the sea.

For any ship, but for smaller ships in particular, the trough can be a dangerous place to be in a storm. It is important, therefore, to avoid getting hung up when a course change takes the ship through the trough. For such a course change it is important to wait for a momentary reduction in sea height, use as much power as conditions permit, and use full or even hard rudder for the turn. Some shiphandlers advocate the use of a twist on twin-screw ships, but this is not nearly as likely to get a ship safely through the trough as is an adequate ahead bell. Even in circumstances that do not permit higher speeds, it can be helpful to use a large but brief ahead bell to start the bow swinging, and while passing through the trough.

As stated by Kotsch and Henderson in their classic *Heavy Weather Guide*, "There can be few if any absolute dictums on handling vessels in heavy weather, because there are so many 'all depends on.' . . . The essence of heavy weather seamanship is anticipation. This means thinking ahead, considering what might happen, planning a course of action and alternatives, and acting before it is too late."[3]

11

AIR OPERATIONS

———◄◦►———

It was 0230 of the midwatch, and USS *Port Royal* (CG 73) was steaming in company with USS *Ronald Reagan* (CVN 76) a few miles south of Oahu. After some exercises earlier in the watch, things had gotten quieter, and the officer of the deck, Lt. Armand Gomez, was providing some instruction to his junior officer of the deck, Ens. Steven Lee. Following some discussion of plane guard duties, Ensign Lee said, "Does it seem to you that the skipper is a little uptight about steaming with a carrier?"

"Yes," replied Lieutenant Gomez, "but for some very good reasons. Carrier operations are extraordinarily demanding, and carriers can be hazardous. Have you ever heard about the collision between *Theodore Roosevelt* and *Leyte Gulf?*"

"No, but I have the feeling I'm about to," said Ensign Lee.

"Don't wise off," instructed Lieutenant Gomez. "You still have plenty to learn."

After checking the radar and both wings of the bridge, and taking a bearing on *Ronald Reagan,* Lieutenant Gomez returned to the center-line pelorus and recounted the story of the collision.

"This all happened during the midwatch on 14 October 1996. *Teddy Roosevelt* and *Leyte Gulf* were steaming in company when they collided. Fortunately, despite an estimated combined closing speed of twenty knots at impact, no one was killed. The damage, however, was reported to be more than $10 million, and the careers of several officers were ruined. So how does something like this happen?

"Weather was not a factor. It was a clear dark night, and the wind and seas were light. *Roosevelt* had concluded air operations for the night, and the only prescheduled event for the midwatch was an engineering drill scheduled for 0200. Shortly after midnight the carrier commenced steering a zigzag submarine-avoidance plan but did not tell *Leyte Gulf* about it until about 0100. In the meantime, the carrier, as officer in tactical command, ordered the cruiser to a station four thousand yards directly astern. About 0130, the carrier commenced unscheduled electrical load shifts that caused power interruptions to some radars and communications circuits. Because of this, the carrier OOD communicated with the cruiser principally by flashing light. When the scheduled engineering drills started shortly after 0200, the carrier OOD advised the cruiser by flashing light only that speed changes could be expected because of the drills."

"Wait a minute," interrupted Ensign Lee, "how do you know the details down to the time at which things happened?"

"Because the senior watch officer at my last command made all of us memorize the story before getting our OOD qualification. Do you want to hear this story or not?"

"Yes, Sir," replied Ensign Lee. "Sorry for interrupting."

"Well, at 0219 the carrier ordered ahead flank. For the next thirty-four minutes, *Leyte Gulf* attempted to maintain its assigned station as the carrier executed several speed changes but did not question the unsignaled maneuvers. At 0244, *Theodore Roosevelt,* as part of the engineering drills, ordered all engines back full. At the same time, *Leyte Gulf* set flight quarters for a helicopter recovery. Neither bridge recognized the imminence of collision until about one minute before it happened. At this time, the carrier ordered all engines ahead flank and the cruiser ordered all engines back full. Although these orders undoubtedly reduced the impact, they were too late to avert the collision."

"How can something like that happen?" asked Ensign Lee. "What went wrong?"

"There was plenty of fault to go around," replied Lieutenant Gomez. "The carrier essentially forgot about the cruiser and focused upon its internal events, only occasionally communicating partial information to the cruiser. While there is no conceivable excuse for the carrier's unannounced backdown with a ship in station directly astern, the cruiser must take a share of the blame. *Theodore Roosevelt* had been executing unannounced speed changes for more than half an hour prior to the collision. Under those circumstances it was past time for *Leyte Gulf* to forget about maintaining station, take a safe course to open the range, and call the captain to the bridge. In addition, setting of flight quarters on the cruiser distracted the bridge team. It's far too easy for this to happen with any evolution: everyone focuses on where the known action is while a hazard closes on them from elsewhere. It's important to organize the bridge team so that a well-qualified individual is, in effect, if not literally, 'on the other side of the bridge' to detect potential dangers before they become imminent hazards. That's why I send you to the starboard wing whenever we are turning to port. Tunnel vision is one of the most dangerous, and yet one of the most avoidable, hazards of going to sea."

Lieutenant Gomez continued. "Miraculously, there were no serious injuries to personnel from the collision. Both ships' commanding officers and the executive officer of the cruiser were found guilty at mast for dereliction of duty and for improperly hazarding a vessel. In the final accident report on the collision, the commander in chief of the Atlantic Fleet concluded that 'the collision was caused by USS *Theodore Roosevelt* backing into *Leyte.*'"

"Wow," said Ensign Lee, "I'd better get another range and bearing to the carrier."

Much of what we do at sea involves air operations: fixed wing and rotary; amphibious, minesweeping, search and rescue, vertical replenishment, and carrier strike operations. Aircraft carriers supplanted battleships as the capital ships of the Navy during World War II, and carriers operating fixed wing aircraft remain the centerpiece today. Increasingly, however, rotary wing and other short take off/short landing aircraft are playing a significant role, operating from both aircraft carriers and a wide range of other warships. Afloat aviation facilities are divided into three classes: aviation ships (CV/CVN), amphibious aviation ships (LPH/LHA/LHD), and air capable ships (all other ships with aviation facilities.)

Operating with Aircraft Carriers

Understanding aircraft carrier operations is important not just to those on board the carrier. It allows those handling the ships in company to anticipate what the carrier will do, and to react accordingly. Launching and landing fixed-wing aircraft from the deck of an aircraft carrier are among the most complex and demanding evolutions of man, and the ability to do so consistently and safely has developed over many decades.

Aircraft carriers are large for a ship but very small for an airfield. To operate fixed-wing aircraft, they have to use catapults for launching aircraft, and arresting gear for landing. Even with these aids it is necessary to create appropriate wind across the deck for both launching and landing aircraft. Thus when operating aircraft an aircraft carrier spends its time running into the wind, and between cycles usually has to run back downwind to relocate. Flight operations begin with a launch. At a planned cycle time, typically one hour and forty-five minutes, the next launch is made, followed immediately by landing planes from the first launch. This cycle continues until the end of flight operations. A large proportion of a carrier's time is devoted to chasing the wind during flight operations and relocating downwind between cycles. The need to spend a lot of time headed into the wind means that carriers need to keep careful track of searoom. A carrier cannot afford to run out of room before completing flight operations. This requires planning far ahead of the immediate evolution. It is also important to keep track of weather to windward. Fog or squalls can present serious difficulties for flight operations.

Navigational considerations normally dictate that the carrier spend no more time heading into the wind than is necessary for air operations. The goal is to steady up on the proper flight course just before the scheduled launch and to turn back down wind as soon as the last plane is on board. Depending on speed and rudder used, the carrier can lose up to five knots in a turn. This can make it desirable to be at a speed greater than that required for flight operations before starting the turn.

The carrier bridge watch needs to keep track of flight course at all times, even when operations are not immediately scheduled. An aircraft emergency or returning aircraft low on fuel may require an immediate turn to flight course. The engineers and ships in company also need to be kept advised of possible speed requirements so that they may manage their engineering plants accordingly.

Ideally, the relative wind for launching aircraft should be directly ahead for the bow catapults and 10 degrees to port for the catapults in the angled deck. For landing we want the relative wind directly down the angled deck. Since it is also important to keep eddies from the island structure and stack gas out of the landing pattern the flight course of choice is usually with the relative wind 10 degrees to port and at the velocity required for the aircraft to be operated. All carriers have a table of required wind conditions for operating each of their embarked aircraft. In the absence of good wind, the carrier may have to operate at high speed when launching and landing aircraft. In light

Figure 11–1. USS *Nimitz* (CVN 68) in a high-speed turn. When aircraft are embarked, a carrier's maneuvers must be much more restricted. *U.S. Navy photo*

wind conditions the desired relative wind direction of 10 degrees to port may not be achievable, and we have to settle for obtaining the required relative wind velocity from directly ahead. High winds on deck can make aircraft handling more difficult, and it is important for the bridge to maintain close liaison with the flight deck. In general relative winds of more than twenty-five knots should be avoided when aircraft are being moved on deck. Ships in company with an aircraft carrier can accurately predict its course for flight operations by doing their own wind calculations.

An aircraft carrier, despite her size, can maneuver in the open ocean with surprising agility. She usually does not. A restriction on carrier maneuverability stems from the need to move aircraft about the deck to correctly position them for operation. The amount of heel generated in a turn could create the hazard of dumping a plane over the side. Thus carriers restrict the amount of heel generated while turning. Some carrier commanding officers specify a maximum amount of rudder to be used during air operations; some specify an angle of heel not to be exceeded. The purpose is the same: to permit the safe handling of unsecured aircraft during the turn. This makes the tactical diameter of carrier turns somewhat unpredictable, and complicates the task of the shiphandler who is maintaining station on the carrier.

Carriers are unpredictable in other ways. Their internal operations are complex, engrossing, and potentially dangerous. This leads to an inward focus, and sometimes to a tendency to forget that there are other ships in company. This can be relatively mild, as in a failure of the carrier to communicate minor course changes as it seeks the wind. It can be actively dangerous, as in the case which began this chapter, of *Theodore Roosevelt* backing into a collision with *Leyte Gulf* without notification to the cruiser.

While this example was particularly egregious, anyone who has spent time at sea in proximity to an aircraft carrier has his or her own sea stories to tell. Eternal vigilance is a requirement for any shiphandler, but particularly when close to aircraft carriers. The prudent rule for any shiphandler operating in proximity to a carrier is to expect the carrier to do the unexpected. Anyone steaming in company with a carrier needs to be a defensive driver. For the aircraft carrier shiphandler the lesson is to remember the ships in company, and never let other tasks detract from the situational awareness necessary to keep the ship safe.

Planeguarding

Helicopters have become the principal plane guards for rescue operations during flight operations. When ships are assigned as plane guards, they now typically are placed in a position to serve as a reference point for landing aircraft. For example, the commanding officer's standing orders set forth in USS *Enterprise* instruction 3505.1G state, "If a plane guard escort is assigned, notify me and position the escort 170 degrees relative to *Enterprise* at three thousand yards during both day and night flight operations."[1] This puts the plane guard ship in a good position to serve as a visual reference for aircraft in the landing pattern. Whenever operating as plane guard ensure there is a copy of the air plan on the bridge, and that the carrier's approach radio channel is on a bridge speaker for monitoring.

When maneuvering in the vicinity of an aircraft carrier it has become customary to consider that there is an "iron box" around the carrier that should not be entered. Typically the forbidden area of the "iron box" extends one thousand yards astern, two thousand yards on either beam, and three thousand yards ahead. Carriers will sometimes promulgate their own desires in this regard. It is also a matter of prudent seamanship while conducting tactical maneuvers to always turn away from the carrier. While this may sometimes be the longer way to a new station, it is almost always the safer way.

Helicopter Operations

Most warships are configured to operate helicopters, and many have helicopter detachments assigned. The helicopter is a major enhancement to the capability of the ship. Helicopter operations have inherent hazards. The shiphandler must be familiar with the requirements for safe helicopter operations. As with all aircraft, takeoffs and landings can be risky, and in-flight emergencies can require immediate landing. Night operations are the most critical, requiring a reduction in intensity compared to daytime operations. Helo operations also involve hazards to ship's personnel, including injury from cargo mishandling, static electrical discharge, and the hazards of the rotors themselves. Because of these hazards a substantial body of safety instructions has been promulgated. Two useful sources are NWP 42 series (*Shipboard Helicopter Operating Procedures*), and NWP 19-1 series (*Navy Search and Rescue Manual*). A detailed checklist should be available on the bridge of any ship involved in helo ops. It is important to observe all of the required safety instructions, including no blowing of tubes when the helo is on or in close proximity to the flight deck, no throwing of anything over the side during helo ops, and removal of all soft headgear during landing and takeoff. Because helicopters build up substantial charges of static electricity, grounding is important in all evolutions, particularly refueling.

Not all helicopter-configured ships can operate all helicopters. As the number of ships capable of conducting flight operations increased the need for a formal standard certification procedure became apparent. The objectives of the certification process are to establish the complete wind and ship motion envelopes for safe aircraft operation and to validate the facilities and equipment for each ship and helicopter combination. The certification takes into account deck dimensions, flight deck obstructions, ship structures, communications, navigation aids, deck marking and lighting, helo rotor diameter, fuselage configurations, maximum gross weight, and other factors. Except for emergencies, ships are not to land or launch helos for which they lack certification. Responsibilities for the certification process are set forth in OPNAV INSTRUCTION 33120.28 series.

Relative wind and ship motion are major considerations for the shiphandler in operating helicopters. Although a helicopter is designed to hover, it can ordinarily generate more lift and carry more load when flying into the wind. In selecting a course for helo operations, several factors need to be considered. The helo should be given a clear upwind approach with minimum turbulence. If the helo is to hover, then it needs a hover position into the wind giving the pilot and hoist operator a clear view of the ship and the deck from which the evolution is to take place. It is also desirable to minimize the ship's pitch and roll. All other things being equal, a good choice of course will put the relative wind about 20 degrees on the port bow with a relative velocity of ten to twenty knots. However, if a helo transfer is to be made to a location forward on the ship, it is preferable to place the relative wind close to the beam.

Most warships, even if not certified to land and launch a particular helicopter, are equipped to provide helicopter in-flight refueling (HIFR) via a Wiggins quick disconnect fueling rig. In this evolution the wind should be placed 30 degrees off the port bow at fifteen knots or more. The helicopter

Figure 11–2. Helicopter in-flight refueling from surface warships has become a routine evolution. *U.S. Navy photo*

hovers over the ship and lowers a hoist. The crew on deck grounds the hoist, then attaches it to the hose. The helo moves to a position alongside and parallel to the ship's course, connects the hose, and takes on fuel. This requires that the ship steer a course into the wind to assist the helo in maintaining station. When topped up with fuel the helo will move back over the ship and lower the hose to the deck crew. It is vital that the fuel provided by the ship be uncontaminated by water or sediment. This usually requires a dedicated JP5 tank with continuous filtration. Most helicopter pilots will request a sample jar of fuel for visual inspection before taking on fuel from a ship with which they have not been regularly operating.

Personnel Transfer

A frequent evolution for all ships is the transfer of personnel using the helo's hoist. This can be done using a litter or stretcher for an ill or injured person or a "horse collar" for a healthy individual. In either case, make sure the person to be transferred is wearing a life jacket. The ship needs to provide a wind from ahead to assist the helo in hovering, but preferably not directly from ahead, so as to minimize turbulence and stack gas. It is desirable to minimize ship's roll and pitch while conducting the transfer. When both the helo pilot and deck crew are ready the hoist or delivery is made, with the deck crew steadying the individual when within reach. Make sure the hoist is grounded, using a discharge wand or grounding line. Ensure that the personnel involved use approved electrical gloves prior to handling the discharge wand or grounding line.

In conducting helo operations, wind is one of the most significant factors to be considered by the shiphandler. Safe launch and recovery wind limitations are covered in NWP3-04.1M and COMDTINST M3710.2 series. There can be substantial differences between the wind as measured by the ship's anemometers and that which actually exists on the flight deck, but the limitations are generally based on the wind as mea-sured at the anemometers. The hazards of helo operations increase sharply as the specified limits are exceeded. Gusting winds are more hazardous than steady winds. It is also important to be aware of the turbulence that can be generated by ship superstructure, stack gases, and rotor wash or jet blast caused by adjacent aircraft.

Launch and recovery wind limits for a given helo are usually presented in diagrammatic form. A sample diagram is illustrated in figure 11–3. Note that the limitations are more stringent for night operations than for day operations.

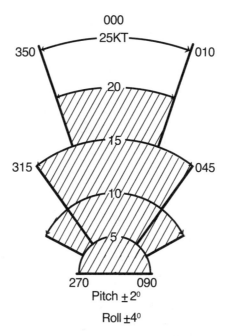

Figure 11–3. Typical helicopter launch and recovery wind limits. Helicopter aligned with ship's lineup line and wind shown relative to the nose of the aircraft. Entire envelope for day operations; shaded area for night operations.

On Deck

The limitations for launch and recovery are not the only concerns. Helo blades are vulnerable to damage when they flap in the wind, so even though the helo may be secure on deck it is important for the shiphandler to keep track of wind conditions. Steady course and speed should be maintained during rotor engagement and disengagement, launch and recovery operations, and any time the helo is being moved on the flight deck. Some helos have blade struts that when installed can provide some extra protection. There are additional wind limitations for engaging and disengaging rotors. Ensure that helicopters parked on the flight deck and in the hangar are properly secured with chocks and tie down chains for anticipated wind and sea conditions. As soon as a helicopter is safely on deck it is tempting to use large amounts of rudder and speed to regain the position lost while conducting helicopter operations. It is incumbent on the shiphandler to keep the helicopter's vulnerabilities in mind before carrying out these maneuvers.

Some ships have special handling equipment to assist in moving the helicopter on deck. An example is the RAST (ship's recovery, assist, securing, and traversing) system as installed on the *Arleigh Burke*–class Flight IIA

Figure 11–4. The RAST recovery system allows helicopters to land on destroyers and cruisers in conditions up to Sea State 5. Here a helicopter lands on USS *Valley Forge* (CG 50). *Ingalls Shipbuilding photo*

guided-missile destroyers. The RAST system is used to move the helicopter into and out of either of the ship's two hangars. It gives the ship the capability of operating SH-60 series helicopters in up to Sea State 5.

Night operations place increased demand on both pilots and deck crews. Safety requires that the tempo of operations be reduced compared with daylight operations. Safe nighttime helo operations require slow, methodical, and careful handling of all aspects of the operation.

Operations with Army Helicopters

Increasingly, Army helos are operating from Navy and Coast Guard decks. This can present additional concerns. Detailed information on Army shipboard helo operations is contained in FM 1-564, *Shipboard Operations*. A major consideration is that Army helos ordinarily use JP4 or JP8 fuel rather than the JP5, which is used by naval aircraft. JP4 and JP8, because of their lower flash points, constitute a fire hazard on board ship. Army helos can use Navy JP5, but their fuel systems must not contain residual JP4 or JP8. Even a small quantity of JP4 or JP8, when mixed with JP5 has an unacceptably low flashpoint. Another consideration in fueling army helos on board ship is that they may not have the needed high-pressure fuel connections. For fire safety the Navy fueling standard is zero tolerance for leaks, seeps, or dripping during refueling. Forehandedness may be necessary to ensure the right equipment is available.

Figure 11–5. U.S. Army troops operate from the deck of USS *Dwight D. Eisenhower* (CVN 69). *U.S. Navy photo*

Figure 11–6. Vertical replenishment adds speed and flexibility to underway replenishment. *U.S. Navy photo*

Vertical Replenishment

Vertical replenishment (VERTREP) is the transfer of cargo from one ship to another by helicopter. It can take place independently, or can augment alongside replenishment. During conventional replenishment VERTREP can speed up the operation, reducing the time that ships must be off station. It can provide support to ships at some distance from the replenishing vessel. VERTREP also provides a means of replenishing units in heavy weather conditions that make alongside replenishment excessively hazardous.

If VERTREP is to be conducted while ships are conducting alongside replenishment, wind considerations must be taken into account in selecting the replenishment course. A relative wind of fifteen to thirty knots from a relative bearing of 330 or 030 is best. The helo will take off, make its approach, and hover into the relative wind. Stack gases or turbulent air from the ship's superstructures can increase the difficulty for the helo.

For VERTREP between ships separated from each other desired wind conditions are the same as for alongside transfers. For transfers at night, it is best for the receiving ship to take station abeam of the delivering ship.

12

TACTICAL MANEUVERING

———◄◦►———

The sun was setting behind Kauai as two Navy frigates steamed leisurely in column formation north of Oahu. USS Reuben James (FFG 57) was the guide at the head of the column, followed by USS Crommelin (FFG-37). Crommelin's officer of the deck, Lt. (jg) Karen Smith, and the junior officer of the deck, Ens. Jason Fremantle, were on the port wing of the bridge, watching as the sun slipped below the horizon.

"It doesn't get much better than this," remarked Lieutenant (jg) Smith. "Do you think we will see the green flash tonight?"

Ensign Fremantle grimaced. "I've been hearing about the green flash ever since I reported on board, but haven't seen it yet. I'm beginning to think that it belongs in the same category as snipe hunts, mail buoys, and left-handed monkey wrenches."

"No, it's real," returned Lieutenant (jg) Smith, "but conditions have to be just right."

Ensign Fremantle looked dubious, but whatever he was going to say was interrupted by the boatswain mate's announcement, "Captain on the Bridge."

Cdr. Sam Ingebritsen climbed into his chair on the starboard side of the bridge and gestured for Lieutenant (jg) Smith to join him. "Karen, we have just received a message saying that a helicopter has gone down about twenty miles to the northeast. *Hopper* (DDG-70) has been directed to join us to conduct a search of the area."

The radio interrupted with a message from *Reuben James:* "Execute to follow, Corpen Starboard 045, over." Ensign Fremantle acknowledged with a "Roger, out." Lieutenant (jg) Smith asked the quartermaster for a stop watch, then reported to the captain, "Skipper, I understand the signal to mean that *Reuben James* is coming right to 045, and we are to maintain station by following in her wake. At our range of one thousand yards and present speed of twelve knots, I intend to hold course for two and one half minutes before putting my rudder over. There is enough daylight left that I should be able to see the guide's rudder kick as a double check."

"Sounds good," said Commander Ingebritsen, just as the radio ordered, "Stand by, Execute." Then, "Execute to follow, Speed 25, over."

As the turn and then the increase in speed were executed, CIC reported an ETA at the crash site in forty-eight minutes. *Hopper* had been identified on radar headed to the same spot.

"That change of speed before I started my turn threw me off a bit, Skipper, but we are almost back on station now," said Lieutenant (jg) Smith.

"Right," said Commander Ingebritsen. "He did throw you a bit of a curve ball, but no harm done. You had better get some extra eyes up on deck to help the search."

About half an hour later, *Hopper* took tactical control of the formation, slowed everyone to twelve knots, came to a course of 045 about a mile ahead of Reuben James, and formed the three ships in a column, with *Hopper* in the lead as guide followed by *Reuben James* and *Crommelin* at two-thousand-yard intervals. As *Hopper* passed through the reported crash site, she ordered, "Execute to follow, Turn Starboard 135."

Lieutenant (jg) Smith looked at the captain. "Looks like she is setting up for an expanding square search, Sir," she said. "When the signal is executed I intend to come to 135 with standard rudder. Our range and true bearing to the guide will remain the same."

"I concur," said Commander Ingebritsen. "My guess is an expanding square to starboard. Ask CIC to set up a plot for that."

Fifteen minutes later the order for a Corpen Sierra to 225 was executed.

Lieutenant (jg) Smith reported the signal to the captain and said, "With a search turn, *Hopper* at the opposite end of the line will turn first. They are already the guide and will remain so. At this speed we should put our rudder over in ten minutes. When we are on station *Hopper* should bear 315 at four thousand yards."

The next ordered search turn was again 90 degrees to the right, to 045. This time Lieutenant (jg) Smith announced, "As soon as the signal is executed we come to 045 and become the guide."

Commander Ingebritsen again concurred, just as combat reported,

"Message coming in saying that report of a helicopter crash was erroneous, Sir. I expect *Hopper* will be releasing us soon to proceed upon duty assigned."

"That is good news, Karen," said the skipper as he got up out of his chair. "Nobody got hurt, and at least it made for an interesting watch. Secure the extra lookouts. I'll be in my cabin reading message traffic."

In most respects, the elements of shiphandling are the same whether for a merchant ship or a man of war. Tactical maneuvering, however, with the partial exception of a merchant ship steaming in a convoy, is unique to naval vessels. Handling a ship in a formation, often at high speeds and in close proximity to other ships, requires both detailed knowledge of the rules for formation maneuver, and the necessary skills to execute the required maneuvers with precision. Most of the requisite knowledge can be found in Allied Tactical Publications ATP-1 and ATP-2. The competent naval shiphandler must know the contents of these publications in detail, for when maneuvering in formation there is rarely time to look up the answer. The maneuvering rules discussed in this chapter must be considered only a preliminary introduction to some of the general principles set forth in these publications, and cannot take the place of

Figure 12–1. U.S. and allied naval vessels steaming in close formation. *U.S. Navy photo*

detailed study. The necessary skills for formation maneuvering need practice to perfect, but are subject to useful principles and rules of thumb that are discussed below.

Rules of the Road

All maritime traffic is subject to the Rules of the Road. The steady growth of maritime commerce since the beginning of the industrial revolution has led to an increasing body of law regulating the conduct of shipping. The early rules were primarily a result of court decisions on maritime accidents. In the latter half of the nineteenth century, and throughout the twentieth, the applicable body of law grew through international treaties. In the last several decades, the International Maritime Organization has assumed the leading role in keeping existing conventions up to date and in developing new conventions as the need arises.

For the shiphandler the most important body of rules are those set forth in the International Collision Regulations, adopted in 1972 and modified at subsequent intervals. The COLREGS are known less formally as the Rules of the Road. They are a uniform set of traffic regulations that apply to both commercial and naval vessels while navigating the high seas. A similar, but not identical, set of Inland Rules govern traffic within the territorial waters of the United States. Every shiphandler must have a detailed and accurate familiarity with the Rules of the Road. Adequately detailed discussion and interpretation of the Rules require book length treatment. A brief summary of some of the more important Rules may be found in Appendix A, but this is only an introduction. The competent shiphandler must know the Rules in detail.

An additional set of rules with which the naval shiphandler must be familiar are those relating to "incidents at sea." One of the most successful international agreements is that which was signed in 1972 between the United States and the Soviet Union, known as the Incidents at Sea Agreement. The agreement specifies that the ships of the respective navies are "to observe strictly the letter and spirit of the International Regulations for Preventing Collisions at Sea." Other provisions include not interfering in the "formations" of the other party, avoiding naval maneuvers in areas of heavy traffic, requiring surveillance ships to maintain a safe distance, using international signals, avoiding simulated attacks, and informing surface vessels when submarines are exercising near them. The agreement also calls for advance notice of certain exercises, and annual meetings to review implementation of the agreement.[1]

In general, the Rules of the Road apply within naval formations, but in some cases they are overruled by special circumstances or by naval practice. The rules discussed here apply to established maneuvering rules between naval vessels, but not necessarily between naval vessels and other ships, except where Rule 2, the "special circumstance" rule, and Rule 27, "Vessels Not Under Command or Restricted in Their Ability to Maneuver" apply. Ships engaged in alongside underway replenishment (not VERTREP), and therefore restricted in their ability to maneuver, have the right of way over other ships, including those engaged in flight operations. Ships engaged in launching or recovering aircraft have the right of way over other naval vessels except for those engaged in alongside replenishment or minesweeping. Helicopters with dipping sonar down have the right of way over all naval vessels, which should approach no closer than five hundred yards.

Some of the rules for maneuvering in formation are based on established naval tradition. Ships in the screen are to keep clear of ships in the main body. Ships not in station should keep clear of ships that are in station. Naval vessels that are not part of a formation should avoid passing through a formation. Small ships should exercise care not to hamper the movements of large vessels, particularly in restricted waters.

Naval vessels maneuvering in close formation are in a sense always in danger of collision. An error can have immediate and tragic consequences. It is therefore vital that all naval shiphandlers know the Rules of the Road, develop the requisite skills, and, perhaps most important, maintain a 360-degree awareness of what is going on around them. It is far too easy for a ship's watch team to concentrate so intensely on what is going on to starboard that they miss the developing situation to port until it is too late. A good naval shiphandler knows what is going on everywhere around him or her, all of the time.

Terms and Definitions

As with any specialized endeavor, naval formation steaming has its own specialized terms with particular meanings. Some of these terms are listed below:

Base course: Intended reference course for the formation.
Base speed: Intended reference speed for the formation.
Column: A formation of ships in a single file line, usually with the forward ship as guide.
Corpen: A formation maneuver in which ships change course in succession, maintaining the same relative bearing to the guide.

Disposition: Any ordered arrangement of two or more formations proceeding together.

Formation: Any ordered arrangement of two or more ships proceeding together.

Guide: The ship on which other ships take station when forming up or maintaining station.

Large ship: Ship length greater than 450 feet.

Line abreast: A formation of ships with each abeam of the other.

Main body: The principal ship or ships of a formation.

Maneuver: Any change of course, speed, formation, or combination thereof requiring ships to adjust or take new positions.

OTC: Officer in tactical command.

Screen: An arrangement of ships and/or ASW helicopters designed to protect the main body.

Small ship: Ship length under 450 feet.

Standard distance: The prescribed range between the foremasts of adjacent ships in a line formation, depending upon size. Normally one thousand yards between large ships, one thousand yards between large ships and small ships, and five hundred yards between small ships.

Standard tactical diameter: The tactical diameter prescribed for use by all ships in a formation so that they will turn together.

Station: The prescribed position of a ship in relation to the guide.

Stationing speed: The speed to be used by ships when maneuvering or changing station.

Tactical diameter: The diameter of the half circle a ship will transcribe when turning 180 degrees from the original course from the point at which a particular rudder angle is ordered.

Turn: A formation change of course in which all ships turn together and retain the same true bearings to the guide.

Stationkeeping

Formation steaming and formation maneuvers are based upon the assumption that ships are in station. Although close formation steaming is not as common as it once was, the accuracy with which a ship keeps her station still contributes to her reputation. Most commanding officers will establish their own rules of thumb as to the limits of station, but a good target is to be within

100 yards of assigned station. Since a station is defined in terms of the range and bearing to the guide, determining the limits of station requires some conversion of range to bearings. If, for example, when on station the guide should bear 270 degrees relative at 3,000, the radian rule (see chapter 9) tells us that 100 yards is equal to about 2 degrees. Thus the limits of bearing and range that lie within 100 yards of the exact station are a range of 2,900 to 3,100 yards and a relative bearing of 268 to 272 degrees true.

Having determined the limits of our station, we then need to measure our position relative to the guide on a continuing basis. Properly calibrated radar is very accurate for range, not as accurate for bearing. The most accurate ranging is by laser rangefinder, and the most accurate bearing by alidade with a gyro repeater. Remember that the station is measured from the foremast of the guide, while most range measuring devices give the range to the closest point on the guide. If the guide is an aircraft carrier and your assigned station is dead ahead, there can be 200 yards difference between the range to the closest point of the carrier's bow and the range to the foremast, and almost that much if your station is astern. As with all shiphandling, it is prudent not to depend on a single source of information. One way of double checking on range is to learn how much of the field of standard Navy 7×50 binoculars is filled by various ship types at different ranges. As a beginning rule of thumb, an aircraft carrier will fill the field of the binoculars at about 450 yards, a cruiser or destroyer at about 350 yards. Unless you are planning to go alongside, either is too close. It is a good idea, when at an accurately known range, to calibrate how much of the field of your binoculars is occupied by different classes of vessels.

Having determined the limits of our station, and how to measure our position, the shiphandler's task is to maintain the ship on station. This requires development of a sense of relative motion, and knowledge of our ship's responses. If our station is directly on the beam of the guide or directly ahead, corrections are easier. For example, if on the beam we can adjust range by alterations to the course being steered, and adjust bearing by alterations in our speed. Changes in course will have negligible affect on bearing, and adjustments in speed will have almost no effect on our range. If our station is dead ahead or dead astern, the opposite is true, with speed affecting range, course affecting bearing. All that is required of the shiphandler in these circumstances is knowing how much of an adjustment to make. Using the earlier example of a station at 3,000 yards on the starboard beam of the guide, assume that we measure our actual distance at 2,900 yards. To move back toward the center point of our assigned station, we need to alter course to the right. How much? Assuming a formation speed of sixteen knots, the radian rule tells us that each 1 degree of course change should alter our distance by about nine yards per minute. Thus in theory a 1-degree course change to the right should bring us back to station in about eleven minutes, with minimal effect on our bearing. In the real world things are rarely this neat. Once we are firmly settled on station, and know what course and speed the guide is really making good, which often will be as much as a degree or two and a knot from the signaled course and speed, we can make adjustments of a degree in course or a few turns in speed. Until we are firmly settled in station, however, somewhat bolder changes are called for. In the case described, a 5-degree alteration of course to starboard should bring us back on station in a little over two minutes.[2]

It is when our station is something other than directly abeam or ahead/astern of the guide that changes in our course or speed affect both range and bearing and adjustments become a little more complex. If, for example, our assigned station is 45 degrees on the starboard bow of the guide, any adjustment of course or speed will affect both our bearing and our range. If we need to open our range, we cannot just alter course a few degrees to starboard, for that will also cause the bearing to change. Thus to maintain our bearing while increasing range we must increase speed somewhat while steering to the right of base course. It does not take long to develop a feel for the relative motion resulting from adjustments to course and speed. While developing this feel it can be helpful to plot the motion vectors on a maneuvering board.

The goal in station keeping is to settle accurately on station, not to pass through it from time to time as adjustments are made. It is easy to develop an oscillation induced by overly aggressive adjustments and inaccurate timing. If we are well off station we need to correct the situation fairly quickly, but not so quickly as to overshoot. Your goal should be to settle in to the point that only minimal adjustments are needed to maintain your station.

The discussion of stationkeeping so far has assumed that the guide is on a steady signaled course and speed. Sometimes, however, the guide will be following a signaled evasive steering zigzag plan. It is essential that your timing be precisely in synchronization with the guide. Even a minute or two out of synch can create substantial problems. Most of the time this will work well. If not, do not be stubborn in following the dictates of the plan. Adjust your course and speed as necessary to maintain the assigned station. This can be aided by close visual observation of the guide, for a course change will usually be more quickly evident by eye than by measuring changes in distance and bearing.

Formations

Naval formations have two basic functions. One is to organize a group of ships so as to facilitate the movement of the body by the officer in tactical command (OTC) from one place to another. The second is to organize the ships for optimum tactical effectiveness. The formations most often encountered are the line of bearing and the circular formation. A special form of the line of bearing, the column or line ahead, is usually considered the most convenient for transit, particularly if the formation is to pass through restricted water. For most tactical purposes, some variation on the circular formation is often chosen, with ships of the main body in the center and screening ships assigned to stations on the perimeter.

During the days of sail and during the ascendancy of the battleship, the line formation was the preferred tactical formation, primarily because it permitted bringing massed firepower to bear during an engagement. During World War II, the circular formation was considered to provide optimum air defense for the main body. With the advent of the threat posed by the kamikazes, and subsequently by guided missiles, the screens were often augmented by air-defense pickets located on the threat axis well away from the main body. For protection of the main body against diesel submarines, whose submerged speed was less than that of the surface formation, the bentline screen forward of the main body was preferred. Tactical formations thus change with the circumstances with which they are designed to cope.

Line Formations

Line formations offer ease of station keeping and maneuvering of the formation. There are several forms of the line formation. Variants include the column, the column open order, the line abreast, the line of bearing, and the diamond formation (see fig. 12–2). The column is a formation of ships in single file, with the leading ship normally serving as the guide. The formation is easy to maneuver and the best for transiting a narrow channel. For open ocean transits, the column open order is preferred. The characteristics of the column open order are that the lead ship is guide, the second ship is 4 degrees off the port quarter of the guide at standard distance, the third ship is 2 degrees off the starboard quarter of the guide at twice the standard distance, and the remaining ships alternate from port to starboard in the wakes of the ships ahead. The column open order has the advantage of allowing all ships to see the others, facilitating visual communications, minimizing radar blind spots,

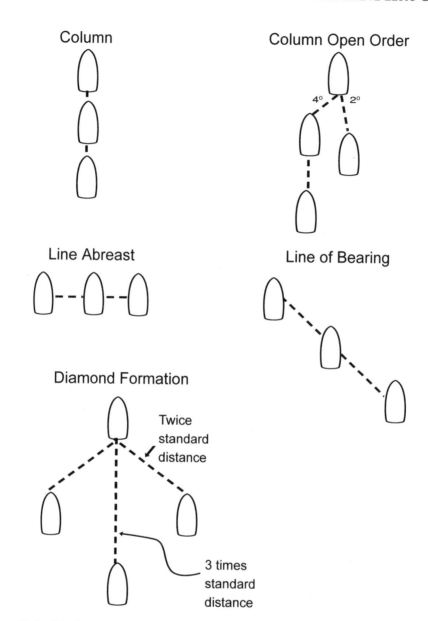

Figure 12–2. Ship formations.

and making for easier station keeping. It adds an element of safety in that all ships are already offset to the side to which they would sheer in case of emergency: "even numbers to port, odd numbers to starboard."

In a line of bearing, ships are formed on the axis of either a true or relative bearing from the guide. A special form of the line of bearing is the line abreast, in which the ships are formed abeam of the guide. This formation is the one most often used in conducting searches, permitting the sweeping of a large area of ocean.

The diamond formation maximizes distance between ships without lengthening the formation. It is often used for the main body of a formation. In the diamond formation the forward ship is normally the guide. The second ship takes station on the guide's port quarter at twice the standard distance, the third ship takes station on the guide's starboard quarter at twice the standard distance. The fourth ship takes station directly astern of the guide at three times the standard distance.

Line Formation Maneuvers

The officer in tactical command maneuvers the formation by signal, which can be visual or by radio. When time permits the signal is usually "execute to follow," which provides a certain amount of time for all units to understand what they are to do upon execution. Not infrequently, however, a signal will be transmitted as "immediate execute." It is therefore vital to understand the meanings of signals even when there is no time to look them up. Some of these standard maneuvers require simultaneous actions by all ships, and some require execution in succession. There are three basic prowords used in tactical maneuvering signals. They are "Form," "Turn," and "Corpen." Each is executed in a distinctive way.

The Form signal is used to change the axis of the formation without changing the formation course. When it is executed, the guide remains on the formation course while the other ships in the formation reorient to form a new line of bearing based on the guide axis, always at standard distance unless otherwise designated. The line of bearing can be either true or relative bearing from the guide. Ships can be ordered to form in order of assigned sequence numbers, or in quickest sequence.

In response to the execution of a Turn signal, all ships turn simultaneously to the new ordered course. True bearings to other ships in the formation remain the same; relative bearings change. The Turn signal can be used to order a turn of a certain number of degrees or to come to a specified course. At night, in low visibility, or when ships with dissimilar turning characteristics are involved, the "Turn" signal is not used for turns of greater than 90 degrees.

In a Corpen maneuver, relative bearings to other ships in the formation remain the same, but true bearings change. A Corpen is executed out of a column, line abreast, or diamond formation (see fig. 12–3). If the formation is steaming in column open order all ships automatically form a column as soon as the signal is understood, and prior to the execution of the signal. The maximum course change by Corpen maneuver is 180 degrees for a column, 90 degrees for a line abreast, and 30 degrees for a diamond formation. For a Corpen maneuver from a column formation the lead ship, as guide, uses standard tactical diameter to come to the new course upon execution of the signal. The other

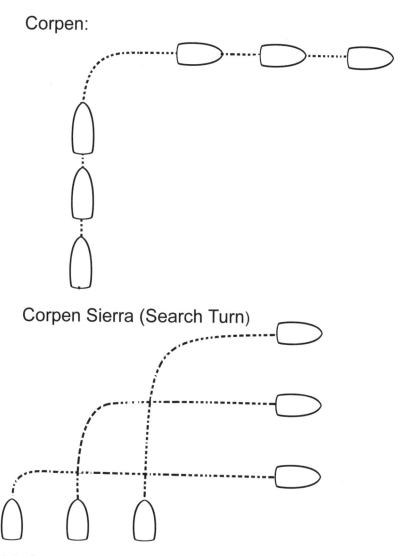

Figure 12–3. Corpen maneuvers.

ships follow in the wake of the guide. For a Corpen maneuver from a line abreast formation the end ship in the direction of the turn comes to the new course upon execution of the signal, becoming the pivot ship and new formation guide. The other ships increase speed to regain the original line abreast formation.

The "search turn" is a special Corpen maneuver carried out from a line abreast. It is used to search a large body of water intensively, to search for a man overboard, accident debris in the water, etc. Ships must be stationed at a standard distance of at least one thousand yards, and the turn must be at least 45 degrees, but no more than 135 degrees. When the signal is executed, the ship at the end of the line abreast *away* from the direction of turn comes to the new course and becomes the guide. Other ships turn in succession to form up on the beam of the new guide at the specified distance. In the search turn you do not have the benefit of a wake to follow, so the stopwatch is the best guide to the right time to turn to wind up at the proper distance and on the beam of the guide. It is also possible to use the maneuvering board to determine what the guide should bear at the time to turn. In practice, the interval between when the search turn is signaled and when it is executed is short enough that only the most adroit users of the maneuvering board are likely to have their solutions.

Other maneuvers from a line formation include interchanging the stations of two ships in the line and inversing the order of ships in a column. To exchange the stations of two specified ships in a line abreast, the ship to port drops back then takes position astern of the ship whose place is being taken. That ship then proceeds to the vacated station. To exchange stations in a column formation, the advance ship hauls out to port and the after ship hauls out to starboard. Both then proceed to their new stations. To inverse the order of ships in a column, upon execution the rear ship in the column hauls out to the designated side, increases speed to one knot less than stationing speed, and becomes the guide. The other ships in column decrease speed to seven knots, then speed up to drop in astern as the new guide progresses up the column.

Circular Formations

The circular formation is based upon a formation center and a formation axis. The formation center is designated as station zero. Any ship in the formation may be designated as the guide, but normally if station zero is occupied that ship serves as the guide. Ships are stationed on concentric circles around station zero, on bearings oriented to the formation axis. Stations to the right of the formation axis are designated by odd numbers, stations to the left by even numbers. The OTC can turn the formation by turn signal, in which case all ships turn together. True bearings remain the same, but relative bearings change. The OTC may also rotate the axis of the formation by up to sixty degrees at a time. When the axis of the formation is rotated relative bearings remain the same, but true bearings change, so that all ships other than the one in station zero must restation.

Screens

Screens are designed to protect the main body of a formation. The screen chosen is based upon the most serious threat. The principal form of screen is circular. Stations may be assigned either at a specific range and bearing from the guide, or as sectors to be patrolled. In a sector screen each screening unit is assigned an area of responsibility, usually defined by inner and outer ranges from the guide and left and right bearings. Areas of responsibility may vary in size to make optimum use of individual unit capabilities.

In a sector screen, ships patrol their sectors, making it more difficult for an opponent to passively determine both individual and formation courses and speeds and the relation of stations to the main body. A sector screen is most effective if the screening units aggressively patrol their areas, changing both courses and speed. To do this requires maintaining an average speed greater than that of the main body, but speed changes are desirable. The screen is most effective if screening units keep track of the movements of their neighbors to avoid leaving exploitable gaps in the screen. Ships should avoid getting closer than five hundred yards to the boundaries between sectors.

Changing Station

The precision and effectiveness of tactical maneuvers depend upon prompt and precise execution. As the old salt says, execution of a signal should be followed instantly by a "puff of black smoke and a swirl of white water." Modern engineering plants have done away with the puff of black smoke, but the expectation of rapid response remains. Much of a ship's reputation rests upon how well she maneuvers in formation.

Upon receipt of a tactical signal requiring maneuver, the first thing to do is to make sure you properly understand the signal, then determine the new range and bearing to the guide. At this point the conning officer should have a good approximation of what he will need to do to get to the new station: come right and increase speed, follow in the wake of the guide, and so on. In the meantime, both bridge and CIC are working to arrive at a more refined maneuvering board solution. If the signal is executed before their calculations are complete, the conning officer is still ready to respond: any necessary adjustments can be made while on the way to station. More importantly, by calculating an approximate solution in his head the conning officer has a quick check on the accuracy of the maneuvering board solution. It keeps you from being vulnerable to someone taking the wrong end of the arrow in calculating the maneuvering board solution. The maneuvering board solution is usually done point to point, and does not take into account turns, either to begin the maneuver or to turn into station. Thus the computed solution can easily be two tactical diameters off. An experienced hand at the maneuvering board can incorporate turns into the solution, but this takes skills and time that may not be available. A good conning officer learns to take the maneuvering board solution for the approximation it usually is, and adjust his path to station to take into account the final turn.

Once you have determined what needs to be done to get to the new station, the next step is to understand whether there are any obstacles between you and the new station, and what the ships in company will be doing. Remember: always turn away from the carrier, and do not get inside the carrier's "iron box," typically one thousand yards astern, two thousand yards on either beam, and three thousand yards ahead. For some maneuvers this will require two legs to get to your new station. So be it. You also need to keep track of where all of the other maneuvering ships in the formation are headed. It is far too easy to concentrate on your own maneuver to the point that you lose track of what others are doing. It is particularly easy to forget about ships astern of you. Bridge resources permitting, it is always wise to have an experienced person assigned to the off side to help stave off tunnel vision. Should your path to station take you close to other ships ensure that they understand your intentions. This is often best done by "signaling with your bow" by exaggerating your heading enough that there is no possibility of ambiguity as to how you will pass. If conditions of electronic emission control permit, it may also be prudent to advise other ships of your intention by voice radio.

Line of Bearing Maneuvers

Other than underway replenishments most formation steaming that involves ships maneuvering in close proximity is done in line of bearing formations.

Columns and line abreast formations are special cases of the line of bearing. Line of bearing maneuvers can be turns, in which all ships in the formation turn simultaneously, or they can be wheels, in which individual ships alter course sequentially.

Turning together is the simpler maneuver: order standard rudder to come to the new ordered course. Always move to the wing of the bridge toward which you are turning. It is not unknown for adjacent ships to turn the wrong way, and when steaming at standard distance you need to be in a position to observe such an error instantly, to prevent a mistake from turning into a tragedy. Ideally you will be precisely on station when the signal is executed. If not, with experience it is possible to adjust station during the turn by using more or less than standard rudder and increasing or decreasing speed a bit. Until experience does provide the necessary touch it is better to perform the maneuver, then adjust station.

Wheeling maneuvers include column movements and search turns. In a column movement all ships follow in the wake of the guide. The key to this is knowing when to put the rudder over to remain in the wake. There are at least three ways to do this, and the prudent shiphandler will make use of all three. Since we want to turn in the wake of the guide we can observe the swirl or kick in the water caused by the guide's rudder. To allow time for the lag between when an order is given and when the rudder responds, the time to turn is about the time the rudder kick is abeam of our bridge. If the water is rough, or if we are several ships back in the formation, the guide's rudder kick will probably not be observable. In this case we can hope that the ship ahead of us is turning properly and turn on his rudder kick, or we can rely on timing our turn. It is an easy matter to calculate how much time should elapse between the execution of the signal and when we should turn. A stopwatch should always be kept handy on the bridge for timing. If the formation speed is fifteen knots, we are moving at five hundred yards per minute. If we are on station two thousand yards behind the guide our rudder order should be given exactly four minutes after the execution. We need not allow for lag time between the order to the helm and the rudder response in this case because the guide will have given the rudder order at the moment of execution. The response lag times will therefore cancel out. Other speeds and other distances will, of course, alter the timing, but the principle is the same.

A third way to determine when to put our rudder over in a column turn involves the use of the maneuvering board. To do this, the guide's projected track upon executing the turn using standard tactical diameter is laid down originating in the center of the maneuvering board. Own ship's course is laid down originating from our station astern of the guide. A table for determining the projected bearing of the guide (assuming a one-thousand-yard standard tactical diameter) follows. (This table may be used to determine the proper point to put your rudder over in a column movement, using a standard tactical diameter of one thousand yards. It is derived from maneuvering board solutions, and so may be in error by a pencil width or so, but should serve as a useful approximation.)

	500 Yards	1,000 Yards	1,500 Yards	2,000 Yards	3,000 Yards	4,000 Yards	5,000 Yards
10°	9°	9°	10°	10°	10°	10°	10°
20°	16°	18°	19°	19°	19°	19°	20°
30°	22°	26°	27°	28°	28°	28°	29°
40°	26°	33°	35°	36°	37°	37°	38°
45°	28°	36°	39°	40°	42°	42°	43°
50°	29°	39°	42°	44°	46°	47°	48°
60°	29°	45°	49°	52°	55°	61°	62°
70°	29°	48°	56°	60°	63°	65°	66°
80°	29°	53°	62°	67°	71°	73°	75°
90°	29°	55°	68°	74°	80°	82°	84°
100°	29°	57°	72°	80°	87°	90°	92°
110°	29°	58°	76°	86°	95°	99°	101°
120°	29°	58°	80°	91°	102°	107°	110°
130°	29°	58°	82°	97°	110°	116°	119°
135°	29°	58°	84°	99°	114°	120°	123°
140°	29°	58°	85°	101°	117°	124°	128°
150°	29°	58°	86°	105°	124°	132°	136°
160°	29°	58°	86°	109°	131°	141°	146°
170°	29°	58°	87°	111°	138°	149°	155°
180°	29°	58°	87°	114°	145°	158°	164°

To use the table, (1) select in the left-hand margin the number of degrees of the ordered course change, (2) move across to column giving the range from your station to the guide, and (3) the number of degrees at the intersection gives the number of degrees of relative bearing the guide should bear to either port or starboard at the point standard rudder should be ordered in order to wind up directly astern of the guide.

In a column movement we want to turn in the wake of the guide, presuming it is visible. We can usually get a visual indication of our turn by watching where the outside edge of the guide's wake lies in relation to our ship's bow (or jackstaff, if it is erected.) If the outer edge of the wake remains in the same position relative to our visual reference, then we are tracking properly through the turn. If the edge of the wake is moving down relative to our reference, we are starting to go wide in the turn. If it is moving up, then we are going to wind up on the inside of the turn. Armed with this information, we can do some adjustment of our turn, but it needs to be done with due regard for other ships in the formation. If we are turning inside and ease our rudder, we will slow less in the turn and may close on the ship ahead. If we are turning wide and wish to tighten our turn, we must be certain not to crowd the ship astern of us. If in doubt, complete the turn with standard rudder, then adjust station as necessary after the column has steadied up.

A corpen movement can also be executed from a line abreast formation. In this case the guide automatically shifts to the end ship on the inside of the turn. The other ships in formation increase speed and maneuver independently to regain their positions abeam of the guide on the new course.

13

SPECIAL SHIP CHARACTERISTICS

———◀◦▶———

USS *Henry W. Tucker* (DD 875) was returning to its home port of Yokosuka, Japan, after a period at sea. It was a cold, blustery day, and the new commanding officer had the conn. This would be his first landing with his new command, but he felt comfortable with the conn, having just come from command of a minesweeper where he had extensive shiphandling experience.

Turning to the XO, the captain said, "I intend to go in with a fair amount of way on to minimize the effect of the wind. I'd like you to stay with me and give me the benefit of any advice you might have."

"Aye aye, Sir," replied the executive officer. "I'll be right here."

With ten knots of way on, the captain made a port turn to line up with the assigned berth. As they approached, the XO said, "It feels a little hot to me, Sir, I recommend that we slow."

The skipper responded with "All engines stop," followed shortly by "All engines back one-third." As the ship continued to close quickly on the pier, he ordered, "All engines back two-thirds," followed by "All engines back full."

The destroyer slowed quickly, and with "All engines stop" settled into position twenty feet opposite the assigned berth. "Over all lines," said the skipper, as the wind moved the ship the remaining few feet to the pier.

As soon as all lines were doubled and the special sea and anchor detail secured, the XO said, "Beautiful landing skipper, but I have to admit that you had me worried for a while."

"Me too, XO. I should know better. Unless there were no alternative, I would never plan a landing that depended on a back full bell. No matter how experienced a shiphandler you may be, it is important to remember that different ships have different characteristics. I didn't take into account that even though we have a lot more horsepower than my minesweeper did, we don't respond as quickly to an engine order. I was lucky that things turned out so well today, but I'll be more careful next time."

Every class of ship has its own idiosyncrasies with which the shiphandler must become familiar. Nothing takes the place of experience in actually handling a ship type, although time on the simulator can come close. Before handling an unfamiliar class of ship, it is helpful to know some of its special characteristics. This chapter examines some of the more numerous classes of ships while trying to minimize duplication of the more general information provided earlier.

Handling the FFG-7

The *Oliver Hazard Perry*–class (FFG-7) frigate was once one of the most numerous in the United States Navy. It is now mostly gone from the active fleet, but many still exist in reserve and in allied navies. A number of them are expected to be used in counter drug operations. The FFG-7 is likely to be around in one status or another for some years to come, and certain unusual design features are of particular interest to the shiphandler.

The distinctive features of the FFG-7 class are two trainable auxiliary propulsion units and a single screw that has a continuous right-hand rotation whether the adjustable-pitch propeller is set for ahead, stop, or back. The screw rotation generates a notable starboard sternwalk under almost all circumstances but particularly with a back bell. The trainable APUs, however, when properly used, make the FFG-7s one of the most controllable of warships.

The basic technique for handling an FFG-7 in close quarters is simple: balance an ahead bell of three to five knots on the engine against an astern thrust on the APUs. The wash of the ahead bell against the rudder permits the stern to be moved to either port or starboard by use of the rudder. The bow can be moved by adjusting the direction of thrust of the APUs. The ship can be moved ahead or astern by adjusting the pitch on the propeller. Thus so long as wind or current do not exceed the thrust vectors that can be generated, the FFG-7 can be placed with great precision without the use of tugs.

In order to lower the APUs to working position, the ship's speed must be five knots or less. It takes about two and a half minutes to lower the APUs, so their use must be anticipated and planned for. The APUs are positioned by ordering the bearing *toward* which you want them to push. Their speed is not controllable; they are either on or off. Because of limitations on how often they can be started and stopped (to avoid overheating), it is normally best while maneuvering to keep them on, and do your adjusting with the ship's engine. Although the APUs are individually trainable, it is easier to keep both trained on the same bearing. This is simpler to keep track of, and avoids their head/tail boundary restrictions.

You can get fancier later, but to begin with, using an ahead bell to balance against the APUs, try using your APUs at 150 to move the bow to starboard, or 200 to move the bow to port. Note that less offset from centerline is needed to move the bow to port because of the starboard sternwalk from the

Figure 13–1. Auxiliary propulsion unit on USS *John A. Moore* (FFG-19). *U.S. Navy photo*

screw. Using the APUs in this way you will find that the ship's pivot point moves somewhat aft. This technique will let you move the ship sideways against a beam wind of up to about twelve knots or the equivalent amount of current. The FFG-7 has a lot of sail area, and above that velocity it is prudent to take a tug or use the ship's single anchor, as discussed in chapter seven. The FFG-7's hull plating is thin and should be treated with care. The sonar dome is located on the keel and does not present a problem either in making up a tug or laying the bow alongside the pier.

The FFG-7s have some other notable characteristics. The combination of the continuously turning shaft and an adjustable-pitch propeller means quick response to engine orders. This quick response dictates that some caution is needed to avoid getting undesired way on while alongside the pier. It is usually possible to hear a change in the sound of the turbines, letting you know that your order is taking effect. When it is necessary to twist the ship, the sternwalk can be used to twist her to port by alternating ahead and astern bells with the rudder left full. A starboard twist is more difficult, but if you must, oppose the engine with the APUs and go ahead with right full rudder.

In almost all maneuvers the FFG-7 is precisely controllable. To back straight, it is usually best to use the engine ahead at three knots with the rudder at right full to balance the sternwalk, relying on the APUs trained at 180 to generate sternway. Once sternway is generated, stop the engine and use the rudder and APU train to control the ship. To make a port side to landing, make a shallow approach, and get line 6 over quickly to control sternwalk. In making a starboard side to landing, come in at a somewhat wider angle, and let sternwalk bring the stern in.

Destroyers and Cruisers

The current generations of destroyers and cruisers in the U.S. Navy, including the *Spruance*, *Ticonderoga*, and *Arleigh Burke* classes, have the same basic configuration: gas turbines, twin screws and rudders, and controllable-pitch propellers (CPP). At lower speeds (below twelve knots for the *Spruance* and *Ticonderoga*, between six and ten knots, depending on plant configuration, for the *Arleigh Burke*), speed is controlled by changing the pitch of the propellers. Above that speed the pitch remains fixed, and speed is adjusted by varying the shaft rpm. Compared to earlier generations of destroyers these are large ships, approximating 10,000 tons. Although powerful and maneuverable, in anything less than optimum conditions their size makes it prudent to use at least one tug when landing or getting under way.

The *Spruance* and *Ticonderoga* share the same basic hull design. The *Arleigh Burke* class is shorter and beamier. A major difference from a shiphandling point of view is the way they react to the wind. The *Ticonderoga* has the most superstructure, and consequently is most sensitive to the wind, to the point that in windy conditions additional care is warranted to avoid being set

Figure 13–2. The current generations of destroyers and cruisers in the U.S. Navy share the same basic configuration: gas turbines, twin screws, and controllable-pitch propellers. *U.S. Navy photo*

outside the channel. This high superstructure makes the ships top heavy, and has led some commanding officers to put limitations on permissible speed/rudder angle combinations in normal steaming to reduce the amount of heel. The rule of thumb most often heard is not to exceed a combination of speed and rudder angle of more than thirty. *Arleigh Burke* is the least sensitive to the wind. All three classes tend to back into the wind, but with the addition of a helicopter hangar to the Flight IIA *Arleigh Burke*s they become more sensitive to the wind, but with more balance between sail area forward and aft. While the earlier ships want to lie stern to the wind, the Flight IIAs tend to lie beam to the wind.

All three classes have large and sensitive bow-mounted sonars, necessitating care when working around the pier. Quick handling of line six is important when coming alongside to keep the bow from swinging into the pier and possibly damaging the dome. The domes are deep enough in the water that they will clear the majority of harbor tugs in use, but if receiving services from a large and unfamiliar tug type it is well to verify the clearance. Because the bow flare can make it difficult for a tug working under the bow, it is often best to place the tug opposite the gun mount on the bow. This has the additional advantage of ensuring clearance with the dome. If a second tug is used, it should be made up on the quarter, rather than all of the way aft.

Sternwalk is not normally a consideration on twin-screw ships, since the screws turn in opposite and canceling directions. On the *Spruance* and *Ticonderoga* classes, the screws as viewed from astern turn inboard when going ahead. The *Arleigh Burke*s have the more conventional outboard turning screws. This is a matter of mostly academic interest, since none of them are notable for their ability to twist. Where sternwalk is a consideration, however, is when one shaft is turning and the other stopped. An operating port shaft, with starboard shaft stopped, will move the stern of the *Spruance*s and *Ticonderoga*s to starboard, while moving the stern of the *Arleigh Burke*s to port. The opposite is true for an operating starboard shaft with the port shaft stopped. The student shiphandler can envision circumstances in which these characteristics can be used to advantage.

The *Spruance* and *Ticonderoga* classes have their normal pivot points at the conventional location just aft of the bridge. The *Arleigh Burke*s' pivot point is further aft, between the stacks. Since the pivot point moves toward the stern when backing anyway, the *Arleigh Burke*s do not steer as well when backing as do the other two classes.

None of these ships twist as well as would be expected from the combination of twin screws, twin rudders, and high power. This is perhaps because of

Figure 13–3. The bow-mounted sonar dome on contemporary cruisers and destroyers dictates care when working around the pier. *Northrop Grumman*

their relatively great length and fairly narrow spacing between the screws. They can, however, be induced to pivot with some alacrity. If space is available ahead to move forward a bit, a two-thirds ahead/one-third back twist puts enough thrust onto the rudders to turn well. If space ahead is limited, put the rudder over full and leave it there, while alternating ahead and back bells, being careful not to gather sternway. The ahead bell will put an effective wash on the rudders, and since the ship is not allowed to gather sternway it is unnecessary to shift the rudder when turning the engines astern. A variant on this is to back one-third on one shaft, while alternating the other between stop and two-thirds ahead, timing the changes to keep the ship from gathering way. In either case, the purpose is to put a strong wash against the rudder while limiting headway.

Twisting the *Arleigh Burke* is made more difficult by the fact that when set for maneuvering combination the ahead and back bells do not balance. With either a one-third or two-thirds twist the ships quickly gather sternway. This can be compensated for by alternating a balanced twist with a larger ahead bell, but this adds to the degree of difficulty. A better solution has been proposed by the commanding officer of one of these ships, Cdr. Terry Mosher. His proposal involves new settings for back one-third and back two-thirds, which reduce the backing power to balance with equivalent ahead bells. Following experimenting to determine the throttle settings that balance, the proper settings are marked on the programmed control lever with pieces of tape. Details of Commander Mosher's proposal can be found in his Professional Note in the September 2003 issue of *Proceedings*.[1]

Because all of these ships combine substantial draft with high power, squat can be a major concern when in shallow water. The *Arleigh Burke* class has been reported to squat as much as twelve feet at twenty-seven knots in fifty feet of water.[2] This is enough to make the difference between safe navigation and grounding in many locations.

As with almost all ships with controllable-pitch propellers, these ships decelerate quickly in response to a stop bell. There are two reasons for this: one is the propeller changes pitch quickly, without having to wait for the shaft to change speed; the other is that when set to zero pitch, the propeller becomes a "barn door" and therefore an effective brake. These characteristics mean that in anchoring it is possible to hold way longer, making the ship less vulnerable to the vagaries of the wind and current. Five knots can be held until the ship is one hundred yards from the anchorage, then all back one-third. The anchor can then be dropped just as the ship passes over the anchorage and begins to gather sternway.

All of these ships are stable and predictable platforms for underway replenishment. As with anchoring, the ship's rapid response means that way can be held longer in making the approach. Using a six-knot speed advantage, the signaled speed can be rung up as the bridge is abeam the stern of most classes of logistics vessels. The *Ticonderoga*s seem to hold their way slightly longer, so a suggested point to drop to signaled speed is as the anchor capstan passes the stern of the replenishment vessel. These ships seem to be somewhat more sensitive to Venturi effect suction while alongside, so that it is normal for them to stay a little wider than usual, with 120 to 140 feet having been recommended by one experienced commanding officer.[3]

Large Hulls

The ships we are calling large hulls here include aircraft carriers, amphibious ships, logistics ships, command ships, and submarine tenders. They range in size from 15,000 to 97,000 tons, and from 520 feet to more than 1,100 feet in length. Most have twin screws, with aircraft carriers having four. Engineering plants are mostly steam: nuclear generated in the case of the CVNs, conventional steam for the others. Some of the newer ships will have gas turbines or turbocharged diesel engines. Shaft horsepower ranges from 22,000 to 280,000.

With this wide a range, what can we usefully say about the large hulls as a group? Perhaps the first thing is that the greater the tonnage, the more inertia and momentum for the shiphandler to take into account. This has several implications. One is that the greater the tonnage, the lower the permissible lateral speed at which you can contact the pier without causing damage. Hull and pier strength do not increase in proportion to the increase in tonnage. If available, Doppler speed indicators can be of considerable value in determining closing rates both to the pier and to objects fore and aft. Second, the larger hulls have a lower power-to-weight ratio than do smaller warships. Even the 280,000 shaft horsepower of an aircraft carrier provides a power to weight ratio that is less than one-third of that of a destroyer or cruiser. The other large hulls have less power in relation to their tonnage. In consequence large hulls accelerate or decelerate more slowly than do the smaller combatants. The greater tonnage also means that mooring lines are less usable in handling the ships, because the greater momentum can more easily exceed the working strength of the lines. For the large hulls, mooring lines become not a maneuvering aid but just a way to hold ships in place once they are put there by other means.

In almost all cases, the large hulls have substantial sail area, making them more sensitive to the wind. This is a factor both in transiting the channel and in maneuvering alongside the pier. In most of the large hulls the conning station is high above the water. The farther the conning officer is from the water, the more difficult it becomes to judge both speed and closing rate.

The consequence of these factors is that the large hulls have more need of tug assistance than do smaller warships, and more tug power is needed to accomplish any given maneuver. Multiple tugs are used to handle large ships, not because two tugs aren't enough to perform all needed maneuvers, but because of the greater power needed to overcome wind, current, and tonnage.

The larger the ship involved, the more the basic technique for going alongside a pier is to make the approach parallel to the pier, and use the tugs to move laterally into position alongside. The ship's engines are used for fore and aft motion, the tugs for pushing or pulling laterally. Getting under way the tugs pull the ship away from the pier laterally, and the ship's engines again provide the fore or aft vector. If a beam wind is holding the ship to the pier, and sufficient tug power is not available to move the ship laterally up wind, the tug or tugs can be used to pull the bow away from the pier, while the ship's engines are used to twist the stern away from the pier.

Figure 13–4. The great tonnage and large sail area of aircraft carriers and other large-hull warships require more tug assistance than do smaller ships. *U.S. Navy photo*

Aircraft Carriers

The aircraft carrier is a very special kind of large hull, with idiosyncrasies of its own. Besides its large size, the aircraft carrier is characterized by substantial overhang, projections from the sides, large sail area, and a narrow, offset superstructure. Because the flight deck obstructs vision from the bridge to the surface in all directions other than to starboard, it is of particular importance to construct a detailed shadow diagram showing distances to the surface in all directions.

Because of the projection of the angled deck to port, and the starboard location of the bridge, carriers go alongside a pier starboard side to unless there is no alternative. For the same reasons the carrier will always do underway replenishment from the port side of a logistics ship. Destroyers have been known to replenish from the port side of an aircraft carrier, but the projection of the angled deck makes this unacceptably hazardous for any circumstance other than an emergency. The sharp outward flare at the stern and bow and the multiple side projections of the carrier present special problems in making up tugs. Tugs will often prefer to pull on a single line rather than making up alongside. A camel can be used between tug and ship, if needed. Carriers also need one or more substantial camels between the ship and the pier, properly secured. These will almost always be taken care of in Navy ports, but not necessarily elsewhere. This is important enough to warrant a double check.

The carrier uses more mooring lines than smaller ships, usually ten lines, sometimes more if conditions dictate. Because of the large sail area and tonnage of the carrier it is more than usually important to ensure that all lines take their share of the load, and that strain is evenly distributed. An additional problem can be that the height of the line-handling stations above the pier creates a sharp down angle for the lines. This means that a large part of the strain is up and down, rather than laterally to hold the ship to the pier. It also subjects the lines to greater strain as the ship moves up and down with the tide. To assist in managing this problem carriers generally have mooring bitts recessed into the sides lower than the chocks on the sponsons. These are unusable until after the ship is already tied up, but can then be used to secure and adjust the moor.

Other ways of avoiding the steep vertical angles for the lines are to run them to the opposite side of the pier, if permitted, and to run breast lines from the portions of the ship further away from the pier. A wide camel can greatly reduce this problem.

Because of the offset position of the island, the line of sight from the bridge to the bow is not dead ahead. Some carriers have mounted special staffs on the starboard side to provide a 000 relative reference for the bridge. In the absence of this, the conning officer needs to select structural points ahead of the bridge that provide a dead ahead reference.

Other Large Hulls

Many of the comments on aircraft carriers apply to the other large hulls as well. A significant difference is that some of the large hulls, particularly amphibious and logistics vessels, can change draft radically depending on their condition of load. This can greatly affect their overall sensitivity to the wind. If the draft changes differently forward and aft, it can also affect the way in which the ship reacts to the wind. It is not unusual for a ship to head into the wind at full load, yet back into the wind at light load.

The draft of large hulls makes them particularly subject to shallow-water effect, with resultant slow rudder response. In shallow channels it is prudent to have a tug join early to be able to assist in making turns if needed. In the absence of a tug, the engines can be used to assist a turn. If help from the engines is needed, it is usually better to select a combination that maximizes screw wash on the rudder. For example, you will generally get more help in the turn from going ahead two-thirds on the screw on the outboard side of the turn while stopping the screw on the inboard side than you will from a one-third/one-third twist.

LSDs in particular have a large superstructure forward, and a beam wind blows the bow strongly to leeward. This not only means that special care must be taken in working around the pier, but it also makes the ships difficult to twist through the wind. Fortunately, the wide set screws of the LSD, together with large rudders, do make for a very maneuverable ship, so long as its special characteristics are taken into account.

Submarines

by Rear Adm. W. J. Holland Jr., USN (Ret.)

Submarines are designed for submerged operations. Surfaced, they are deceptively large, heavy for their physical size, and generally unwieldy. Because of their small above water structure, other ships usually do not appreciate their size or deep draft, may even ignore them, and often maneuver with little regard to their presence. When in the vicinity of a surface ship, wherever possible, a submarine should maneuver early to avoid becoming the stand on vessel in a crossing situation.

A submarine's configuration means that even small injuries to the hull can have serious consequences. The extremities of the stern planes contain sonar array tubes that can crush if hit hard. The sonar dome on all submarines and the sides are covered by an acoustic coating and some have sonar hydrophones that are sensitive to the contour and structure of the coverings—even a small dent or dimple can result in damage to the coating or a loss of sensitivity to the sonar array. To a lesser extent, the stowage tube for the towed array is located on the starboard side several feet above the waterline. Many a submarine has cracked the fiberglass fairing covers with poorly managed tugs alongside. Care should be taken to maneuver to avoid damage to these surfaces. Additionally, the placement of cleats and the short forecastle complicate line handling and tug makeup.

The nature of submarine operations limits the opportunities for maneuvering on the surface, in confined waters, and alongside a pier. Handling the ship in these circumstances can even be classified as a rare event—worthy of careful preparation and attention to detail.

Current has great effect on a submarine because most of the ship is under water. Planning to get under way or moor during slack water is always the

Figure 13–5. Submarines are unwieldy around the pier and will almost always take tug assistance. *U.S. Navy photo*

best course. Maneuvering a submarine alongside with a current running adds unnecessary risk to an already challenging evolution.

Wind is rarely a factor in maneuvering a submarine, except as it hinders the performance of topside personnel.

Pilots accustomed to merchant ships or who have no experience with submarines tend to fail to appreciate the momentum of the ship. They also may underestimate the responsiveness and power associated with full bells. These characteristics should be discussed when the pilot boards.

Getting a Submarine Under Way and Landing

Submarines require deep-draft camels or fenders and are designed to have lines provided from the pier. With lines homed on the pier rather than on board, using lines as springs, while not impossible, is not very practical. Deep-draft camels must also be provided when mooring submarines in a nest or alongside the tender. When alongside another submarine of the same class, these camels or fenders must be wide enough to maintain clearance between stern planes of adjacent submarines. These planes are invisible, well below the waterline, and extend beyond the line of the hull. Care must be exercised to prevent these planes from contacting hard objects, that is, tugs, piers, hulls, or other planes.

In other than CONUS and ports regularly visited by submarines, deep-draft camels or fenders may not be provided. When they are not, care must be taken to nestle the submarine to the pier gently to avoid damage to both submarine and pier. Mooring lines must be made up to allow the ship to work gently alongside with the tide and currents yet taut enough to prevent the ship from gaining significant momentum during heavy weather or in response to a large wake.

Deployed operations may require the submarine to carry mooring lines. In these circumstances, when the line is tended from the ship, the ship may use spring lines to swing into the pier. However, unlike other ships that use the hull-pier contact as a fulcrum to lever the ship in, submarines must keep the ship oriented parallel to the pier, avoiding contact if possible during this evolution.

Submarine Secondary Propulsion Motors (SPM)

Attack submarines (SSNs) are equipped with a retractable and trainable motor that should always be used for getting under way and landing. While well aft of the pivot point, this secondary propulsion motor (SPM) has enough power (325 kw) to move the ship laterally off or onto the pier and should be used to put the initial momentum on the ship. Clearing the pier/moor with the secondary propulsion motor can be tedious but it is generally the surest method of clearing the berth when tugs are not at hand.

Ballistic missile submarines (SSBNs) have two auxiliary motors but these are not trainable. Their location well aft limits their utility to maneuver the ship to ahead and astern; they are too close together to add a twisting moment. On the other hand the sides of the ballistic missile submarines are not as susceptible to injury as attack submarines because they do not have wide aperture arrays. Their large and vulnerable bow domes however make twisting alongside the pier using lines undesirable.

The secondary propulsion motor may be rigged out only when the submarine's speed is below ten knots.

Maneuvering Alongside

Tugs should be used getting under way and to maneuver the ship into its final moor alongside. They should be kept made up until the lines are doubled and secured.

The direction of swing on a backing bell is unpredictable. When backing down, the ship may respond to the rudder once sternway has been achieved, but if the previous swing has been vigorous, stopping the swing with the rudder takes more time and sea room than are likely to be available.

Care must be taken when alongside to protect the sonar dome, the wide aperture array coverings, and the stern planes from damage. The desired course is to maneuver to avoid touching these surfaces. Submarines always use tugs to get under way and to moor. Maneuvering in confined spaces without tugs when they are available is foolish (see fig. 13–6).[4]

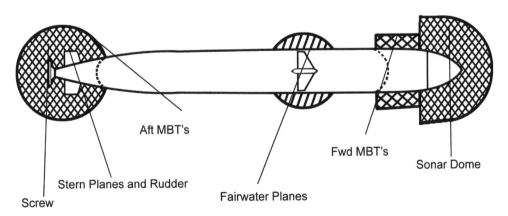

Figure 13–6. Submarine damage-prone areas.

Tugs are normally handled by the docking pilot and while the submarine has options about taking the harbor or bar pilots, the docking pilot should always be on board to advise the captain and conning officer and to control the tugs. In the greeting conference between the commanding officer and the pilot, the tug makeup to be used and the method of controlling the tugs should be established overtly and agreed upon by both parties; for example, the pilot to relay the conning officer's or captain's orders to the tugs or to give orders to the tugs as he deems appropriate.

The preferred makeup for tugs is the full power moor (see fig. 13–7). This makeup, two tugs with three lines each with the forward tug in a reverse power moor, allows the two tugs to walk the ship in either direction and spin it within the ship's length. If only a single tug is available, the combination of the tug forward in a reverse power moor and the ship's engine can move the submarine sideways with good control.

In submarines equipped with fairwater or sailplanes, the potential exists for the planes and the superstructure to interfere with the tug's upper works. In these circumstances, the modified power moor is preferred (see fig. 13–8). In this makeup, the forward tug must be cast off before headway is put on to avoid a collision between the planes and tug superstructure. This moor is less of a threat to the sonar dome than is the full power makeup. Headline and quarterline is a form of this makeup in which the forward tug uses its center-line fairlead for a single line rather than the port and starboard bow fairleads on the bow. Though less secure than the modified power moor, this arrangement is much better than the single headline.

In U.S.-controlled ports, tugs will be equipped with deep-draft fenders, allowing the tug to makeup without riding up on the hull and permitting the

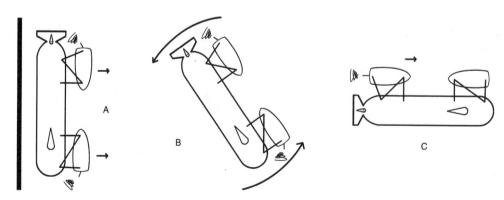

Figure 13–7. Full-power moor.

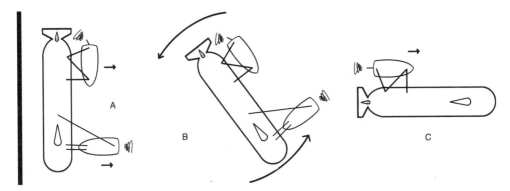

Figure 13–8. Modified-power moor.

use of the power makeups described above. In foreign situations, where tugs have shallow draft or poor fenders, headlines and towlines may be all that can be used because of the potential for hull damage. In these cases, tugs should be made up to pull instead of push; a double headline makeup is the best arrangement (see fig. 13–9).

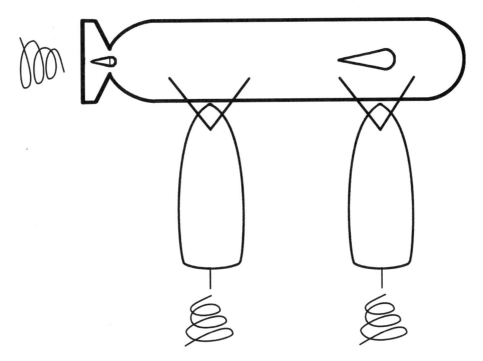

Figure 13–9. Double headline makeup.

Makeups using a single headline or a single towline are very limited in their utility, sluggish in response, and carry the risk of hitting or rubbing the hull hard. These makeups are generally relegated to pulling the ship's head around or towing the ship away from the pier.

Submarine Ground Tackle

Anchoring a submarine is not a desirable evolution and should be avoided if possible. The anchor is a mushroom tucked into a cavity just forward of the rudder. Dropping it in a desired position is problematical and recovery hard.

The anchor should be let go with bare steerageway on the ship—five knots is likely to break the weak link leaving the anchor and chain on the bottom. The dropped position must be recorded carefully in order to facilitate weighing. When weighing anchor, a strain should be put on the chain and the submarine backed slowly as the chain is brought in until the ship is over the drop point. There the main engine should be operated at very slow revolutions to maintain the submarine above the drop point. When the anchor weighs, the stern will begin to swing. The engine is then stopped and anchor brought in. If necessary to maneuver in a seaway after the anchor is aweigh but before the anchor is housed, the engine may be ordered ahead. But no more than bare steerageway should be put on the ship until the anchor is housed. Weighing anchor is the one evolution where the secondary propulsion motor should remain housed. Backing over the chain with the motor extended can damage both chain and motor.

Mooring Alongside

Overseas, when pier space is not available, a submarine normally moors alongside a tender (AS), fleet tug (ATF), salvage ship (ARS), or similar ship that has ground tackle with more than adequate holding power for both ships. Where this moored-to ship is moored both forward and aft, the approach and landing are as at any pier. However, if the ship has ground tackle down only forward and swings with wind and current, the approach and moor are somewhat more complicated.

If there is little wind and a steady current, the moored-to ship will remain fairly steady but if the wind is shifting or the current variable, the mooring ship will be moving during the approach. The submarine should approach into the wind, aiming for a final separation of about thirty feet. The approach should be not be too flat: a 15- to 20-degree angle is appropriate, because the screws of the moored-to ship probably will project beyond its side at the stern

Figure 13–10. When pier space is not available, a submarine normally moors alongside a tender (AS), fleet tug (ATF), or salvage ship (ARS). *U.S. Navy photo*

and the moor must be made to avoid sweeping the submarine hull underneath the moored-to ship's tumble home and striking its screw.

The approach should be as slow as possible in order to reduce the wash from the astern bell when alongside. Too much wash will push the moored-to ship (which usually will be much lighter than the submarine) away, complicating passing the lines. The moored-to ship should provide the lines. Once the bow line is over and down and the submarine is dead in the water, the moored-to ship can close the gap using its outboard screw. The submarine can aid this maneuver using the secondary propulsion motor to close the gap. The main engine may be used gently to maintain the submarine dead in the water fore and aft position but care must be taken. Too much power will push the moored-to ship away and care must be taken not to put way on the two ships lest the moored-to ship be wound around its anchor.

If the moor is missed, the separation too wide or the moored-to ship moving too fast or erratically to be comfortable, maneuvering clear for a new approach is more seamanlike than trying to salvage a deteriorating situation.

The final makeup will depend upon the arrangements of the moored-to ship but will generally parallel the normal six-line arrangement. Where the moored-to ship is shorter than the submarine, the bows of both ships should be approximately even with the stern line leading forward from the submarine through the moored-to ship's stern chock.

When getting under way, "breaking the moor," the moored-to ship should go ahead slowly on its outboard screw taking care not to gain significant headway or overrun its moor. The submarine should use the secondary propulsion motor at sixty degrees off centerline to move away from the moored-to ship. The wind will tend to push the bows of the two ships together so that once started, propulsion needs to be maintained on both ships to keep them moving apart. Once clear, backing the main engine smartly will break the moor before the moored-to ship swings back. However, putting a large backing bell on too soon will cause the bow of the moored-to ship to swing into the submarine.

The only peculiar condition when transiting the channel is to have due regard for the presence of people topside. The submarine's way must be limited when hands are on deck.

Emergencies

Man overboard is likely only in confined waters when personnel are topside for personnel transfer or preparations for mooring. If appropriate precautions, that is, safety harnesses, are taken on the bridge and topside, the likelihood of man overboard is significant only when transferring personnel or when entering or leaving port. Hands should not be brought on deck until in sheltered waters with a tug or a small craft paralleling the ship. During the transfer evolution, the ship should be slowed to bare steerageway, and then, if a man is lost in the water, the shaft should be stopped, the stern swung away from the man in the water, and the recovery performed by the small craft/tug.

In open water or when without an escort vessel, the selection of the appropriate recovery method for a submarine is the same as those discussed in chapter 10.

Coast Guard Cutters

by Lt. Cdr. Joe DiRenzo III, USCG

The United States Coast Guard operates a fleet of cutters, from icebreakers to patrol boats, in some of the most demanding shiphandling conditions imaginable. The Coast Guard faces an increased emphasis on Homeland Security, together with the ever present need to perform search and rescue. Successful execution of these missions requires an understanding of the special shiphandling characteristics of cutters. From underway replenishment to helicopter operations and towing, Coast Guard cutters are involved in many missions, with each class of cutter presenting a unique challenge.

This section looks first at larger cutters, those above 175 feet, and then smaller cutters, such as patrol boats. Not included in this discussion are Coast Guard river tenders, utility boats, rigid hull inflatables, MSBs, or port security vessels.

High- and Medium-Endurance Cutters

The 378-foot *Hamilton*-class WHEC is one of the Coast Guard's most capable platforms, with a forty-two-foot beam, two shafts, controllable-pitch propellers, and a trainable bow prop. The maneuverability afforded by the bow prop gives great latitude in deciding how to moor or unmoor. WHECs do not normally require the use of tugs unless wind and/or current exceed the bow prop's available power.

The combination of twin screws and the bow prop allow the WHECs to "walk sideways." To accomplish this, the engines are opposed and balanced to keep headway off the ship and the rudder is placed to move the stern in the desired direction. The bow prop is then used to move the bow in the same direction, allowing the bow to "keep pace" with the stern's lateral movement. For example, to move the WHEC laterally to starboard, you would use starboard engine ahead one-third, port engine back one-third, and rudder left full, while simultaneously using the bow prop to move the bow to starboard. It becomes a dual balancing act, keeping the engines balanced to avoid headway or sternway and keeping the ship parallel to the pier by balancing the lateral thrust from the bow prop against the lateral thrust from the twist and rudder. In this way the WHEC can be maneuvered with great precision.

Besides the ability to walk sideways, the high-endurance cutter's propulsion system is also unique within the Coast Guard in providing a combined diesel and gas turbine (CODAG) engineering plant. Diesels are normally employed during most shiphandling evolutions such as mooring, unmooring, and towing. Underway replenishment, on the other hand, is usually accomplished on turbines.

Medium-endurance cutters, or WMECs (270 feet and 210 feet, respectively), do not have a bow prop and must be handled differently alongside the

Figure 13–11. *Hamilton*-class cutter USCGC *Gallatin* (WHEC 721). *USCG photo*

pier. These cutters require the conning officer to master the art of "aspect" shiphandling, or as some have called it, "driving the bow." To do this, the position of the bow becomes the primary focus. The conning officer drives the bow to a predetermined position, then "pins" it in place using mooring lines and/or the anchor. The stern is then positioned using engines and rudders. The technique is similar to that used with any naval twin-screw ship, although the WMEC shiphandler has to deal with less momentum than larger naval vessels.

In handling a WMEC, the conning officer must learn to change the aspect of the ship departing or approaching a pier to be able to control the landing. The "aspect" can be changed in a couple of ways. Most obviously, and seemingly intuitive to the novice shiphandler, the twin-screwed WMEC can be twisted left or right by opposing the shafts, one forward and one astern. However, the turning moment between the two shafts is actually quite small, and twisting the ship by doing nothing other than opposing shafts is an inefficient way to change the aspect. Neither the 378 or the 270 twist very well. Of the three types of cutters discussed thus far (378, 270, and 210), it is the 210 that has the greatest horsepower-to-weight ratio. Consequently, this class is significantly more responsive to engine commands than either the 378 or 270 and is the best at twisting. Given constraints, such as limited maneuvering room, twisting is a viable and commonly used option by all cutters, but

Figure 13–12. The *Bear*-class medium-endurance cutter USCGC *Spencer* (WMEC 905). *USCG photo*

the most effective way to change the aspect of a WMEC is to force a lot of thrust across a full rudder. Whether the rudder is left or right full depends upon which direction one wants the stern to go. A strong ahead bell across a full rudder while the ship is stopped, say, an ahead bell left on momentarily so that the momentum does not turn into speed (similar to a tugboat giving a strong bell and immediately taking it off), will quickly change the aspect of the ship. The conning officer needs to understand that with this maneuver there may be some challenges. First, putting a lot of ahead thrust across the rudders will tend to affect the cutter by starting to move ahead, slowly at first, but then more and more quickly. The second issue with this maneuver is that the stern will initially move rapidly in the desired direction since the pivot point is initially well forward. Once headway is gained, however, the pivot point moves aft, and this advantage is lost.

The shiphandler must prevent the ship from moving ahead or do what is sometimes called "pinning the ship in place" or as older salts would advise "backing and filling" by quickly going to a backing bell on the inboard shaft to pin the ship in place or prevent it from coming ahead. By pinning the ship in place the stern can be moved left or right thus changing the aspect of the ship enabling the shiphandler to move her forward or aft away from danger or placing her parallel to the pier. In simplest terms this means that you want to make sure you control your engines to avoid getting way on the ship.

Another option in this situation is to "pump the rudder" by using full rudder and a series of ahead bells, almost immediately taken off, while watching the swing of the stern against the background. You never want to completely stop the swing in the desired direction, because it takes a tremendous amount of momentum to get it moving that way again and shiphandlers tend to overcorrect and apply too much power, for too long, to get the stern moving again, and thus gathering undesired headway.

A valuable tip for maneuvering alongside, with a WHEC or WMEC, is to set up a condition where the outboard shaft is always kept with an ahead bell on and fore and aft motion is then controlled by backing or stopping the inboard shaft. By keeping the ahead bell on the outboard shaft there is always a constant flow of water being forced over the rudder and the cutter will respond to the rudder command in a predictable manner. Similar to keeping the swing of the stern going in the desired direction with a twist, the key is to keep slight headway on until in the desired mooring position. Coming to a stop or gaining sternway negates the control of the rudders and makes the ship more vulnerable to any wind or current effects. A surge of power is then required to regain headway, which can easily turn into excessive speed. As a rule of thumb, a back bell of one handle position greater than the ahead bell (i.e., port ahead 2, starboard back 3) will keep a WHEC or WMEC essentially stationary without creating a twisting motion.

Another way of changing the aspect of the cutter is to use the tendency of all of these ships to back into the wind. WMECs and WHECs will all weathervane away from the wind, meaning that the stern will tend to be "pinned" while the bow, with no force under it, falls off or is blown off until the ship is pointing directly downwind. Actually, the stern simply tends to back and the bow falls off due to the force of the wind on the side of the ship; there is nothing to "pin" the bow like the stern with astern propulsion operating. So in the end, the stern is backing directly into the wind and the bow is following along nicely with not much inclination to fall left or right of the wind line. Weathervaning is a critical factor in WMEC shiphandling. It is often advantageous to "back and fill" with a WMEC, particularly a 270-foot WMEC, rather than trying to power the bow through the wind. Once the bow does pass through the wind, though, it falls off rapidly and the conning officer must always be alert for set and drift while doing this time-consuming maneuver.

Another common shiphandling characteristic of the 378-foot WHEC that must be addressed is when to move to a bridge wing for conning alongside a pier. The WHEC has a forty-two-foot beam, and conning officers can get into trouble by moving out onto the bridge wing too soon. If a conning officer moves out onto the bridge wing early, he will, owing to parallax, get a distorted sense of the cutter's movement during the approach. This problem is less with the WMECs because of their narrower beam. For the WHECs the best approach is to remain at or near the centerline until on "short final" approach to the pier.

The effect of wind has been discussed earlier. WMECs, especially 210s, have a significant sail area and shallow draft (ten to eleven feet). This means they are more susceptible than WHECs to wind effects when mooring and unmooring. Some 210s have had a heel of seven degrees at the pier with a light wind and slack water!

Three final notes on high- and medium-endurance cutters. Each of these issues has sparked significant discussion within the cutter fleet. First, some seasoned shiphandlers believe it is generally best to declutch a 378 at the pier as soon as alongside and the lines are passed and made fast. This would entail using the bow prop to move the ship up or down the pier or to move the bow on or off. The reasoning is that occasionally pitch goes out of control on controllable-pitch ships and that it is best to remove that factor as soon as possible. This approach is also advocated because the pitch cards may not be exactly zeroed and you could unintentionally pick up some unwanted headway or sternway while you are in contact with the pier. However, not all shiphandlers believe this is the case. They assert that a pitch-control failure is probably no more likely than not being able to clutch back in should you need the engines again. By taking away your engines you almost eliminate your rudder. The debate goes on.

Second, both the 378s and the 210s have a lot of sheer to the bow that tends to get blown onto the pier and can knock down lamp posts and the like. The 210s need to have their stern pinned alongside (by twisting) until the lines are set to keep the bow from blowing onto the pier with an on the dock wind.

Third, all shiphandlers on board high- and medium-endurance cutters need to remember the use of the anchor, especially for mooring. This "tool of the trade" is especially useful in the moderate- to high-wind situations which might take charge of a 210's bow.

Buoy Tenders

Unlike high- and medium-endurance cutters, which require shiphandling skills very similar to those used by any Navy counterpart, the special characteristics of buoy tender shiphandling entail a completely different skill set.

Before getting into the specific characteristics, a short review of some of the special systems onboard the 225-foot seagoing buoy tenders (the *Juniper* class) or WLBs (coastal buoy tenders—the "Keeper" class or WLMs) might be of interest.

The dynamic positioning system (DPS) is found on both the new WLBs and WLMs. The DPS is part of both vessels' integrated ship control system (ISCS) which provides the capability to automatically approach and maintain position to within approximately thirty feet of a designated point. This system provides the conning officer the ability to keep a steady heading while hovering over a specific point or the cutter can pivot about a point, such as a navigation aids intended position. Even more unusual is DPS's capability to maintain a position in winds up to thirty knots and seas to eight feet. Along with DPS, the WLMs have the 360-degree Z-drives. The *Keeper* class has no rudders but uses the Z-drive, which pulses water, via a thruster, to effect the movement of the ship. To quote one former operations officer, "These cutters will turn on a dime with barely a wake."

Discussion of the steps involved in "working" a buoy provides the best opportunity to consider their unique shiphandling characteristics. Typically, when beginning, the conning officer will line up heading into the wind or current and engage the "hold heading" feature on the DPS. This system enables the conning officer to use a combination joystick/heading knob or "buttons" to control the ship's movement. For example, one holds the heading on these classes of cutters by using the thrusters (bow thruster on the 175 and bow and stern thrusters on the 225) to maintain the head.

When actually working a buoy, two approaches to shiphandling are used. First, after bringing the aid on deck, securing the chain, taking out slack, and getting the sinker off the bottom, the conning officer engages the "hold position" feature. This gives the DPS control of the thrusters and main propulsion (twin Z-drives on the 175 and one CPP on the 225). Taking DGPS through a fiber-optic LAN, the DPS maintains station. While the crew is working the aid, the conning officer can input commands to the DPS to move the cutter to the aid's assigned position. The conning officer will most usually try to head into the wind or current to reduce the load on the thrusters.

Another option, especially in any type of a sea state, is to approach the buoy and bring it to a position about five to ten feet from the buoy port and then engage the hold position. This allows the ship to settle out its position using DPS. The conning officer can then offset incrementally toward the aid, creating a more controlled evolution with less chance of doing damage to the buoy.

Figure 13–13. The crew of the Coast Guard cutter *Redwood* (WLM 685) prepares to set a buoy. *USCG photo*

Handling alongside also highlights the unique handling characteristics of the 225s. Because of the controllable-pitch propeller, the 225 does not hold its way with a stop bell; the prop acts like a brake. With no way on and the rudder amidships, the stern slowly moves to starboard due to the walking effect of the CPP. Using between 10 and 20 percent on the stern thruster to port offsets the walking effect. Heavy wind on the port quarter makes coming alongside even more challenging. The reason is that the thruster is already working to port to offset the screw so the conning officer must compensate. The easiest thing to do is place the cutter parallel to the pier and walk sideways with the thrusters.

The 225 is single screw with a controllable-pitch propeller and bow and stern thrusters. The 175 has a bow thruster and two azimuthing stern thrusters (Z-drives). Both cutters can spin in circles in their own length. The 175 acts like a jet ski during high-speed turns and actually slides through the water rather than turning. The thrusters are pushing the stern, not turning the bow, since there is no rudder.

Patrol Boats

The Coast Guard's patrol boat fleet, comprising of 110-foot (WPB) and 87-foot (CPB, or coastal patrol boat) vessels, are the service's workhorses. Stationed from Portland, Maine, to Key West to Guam, patrol boats, with two shafts, two five-blade fixed-pitch propellers, and two rudders, are used in every possible mission area. There are, however, some very specific shiphandling characteristics that the prudent conning officer needs to recognize. These can be placed in three categories; along side the pier operations, small boat launches, and anchoring.

With the 110s, maneuvering while alongside the pier is done by clutching and declutching the two main diesel engines. The *A* and *B* class clutch in at nine knots, while the *C* class clutches in at seven knots. To clutch in means the point at which the engines actually engage the reduction gear and the resulting speed generated. This means that the minimum bell available is a strong one. The conning officer must understand early on the amount of force that is quickly developed while clutching engines in and out alongside a pier.

The *C* class, however, does provide another capability, making shiphandling along the pier easier. This is provided in the form of a "troll mode," which provides a speed under two knots, greatly improving alongside maneuvering. Troll mode is accomplished on the *C* class by allowing the clutch plates to slip. All three classes present issues during a tow and can easily get too much speed, especially in the *A* and *B* classes.

The second special characteristic of which the conning officer needs to be very aware involves launching the rigid hull inflatable boat (RHIB). On the 110s, the cutter must be transiting "down swell" with speeds under four to six

Figure 13–14. The *Island*-class patrol boat *Matagorda* (WPB 1303). *Bollinger Shipyard*

knots (different for *A* and *B* vice *C*). There must be constant communication between the deck supervisor and the bridge as the RHIB initially enters the water. More speed than needed, or a misevaluation of seas, can cause the RHIB's bow to submerge. One common shiphandling mistake some new conning officers make is to apply more power, which only exacerbates the problem.

Anchoring a 110 is also challenging because the vessel is equipped, like the 87-foot CPB (discussed below), with an anchor line, rather than an anchor chain. This class of cutter uses a two-inch double-braid line, with the exception of the 110-foot *A* class, which has a partial chain. When anchoring, the conning officer needs to ensure that the anchor is truly dug in and then set chafing gear. There is the potential for a line to part when anchored. The lack of a chain reduces the holding power of the anchor, since the weight of the anchor chain is part of the equation determining holding power. The rode does not have the weight of a chain, so the flukes really need to be set well. The anchor is designed to set horizontally and break out vertically, so you want plenty of rode out before setting the anchor in order to make the pull as horizontal as possible.

One final characteristic that needs to be mentioned regarding the 110-foot WPB is its capability to twist. The 110s have great horsepower-to-weight ratio and a nice lateral separation of the shafts making it possible to twist this class of vessel nearly in place.

The 87-foot CPB is maneuverable at moderate speeds, but the engines take about three seconds to engage at seven knots, which takes some time to understand and anticipate. The large sail area of the superstructure can easily take charge in high-wind conditions. The conning officer must be aware of the prevailing weather when conducting a man overboard, towing, or operating along side a pier.

Unlike the 110, which uses a single-arm davit, the CPB has a stern launched small boat, which is limited to six-foot or less seas. In an ideal world, all crew board the boat, lift the stern gate, and go to sea. You must run with the seas slightly off the bow when launching or recovering. Following seas tend to surf the small boat into the notch, thus introducing more power than the small boat can counter.

Icebreakers

The Coast Guard has four icebreakers, each with unique characteristics. The *Healey*, the service's newest, has twin screws and displaces 17,700 tons. The two Polar-class vessels, which will be our focus, are classic icebreaking hulls with no hard underbody edges at all. The final icebreaker is the *Mackinaw*, one of the service's oldest vessels. The Polar class, with their three shafts, are homeported in Seattle, Washington, and are designed so that the cutter rides up over the ice and then, using the significant displacement, hammers the ship back down, crushing rather than shouldering aside the ice. A single rudder behind the centerline shaft keeps this most important system partially protected but does reduce maneuverability if you lose the centerline shaft. In an open seaway, the roll is excessive, with 50 degrees not uncommon in rough weather. In ice, this is not a factor. With three turbines on the line, ships of the Polar class are prodigious icebreakers, capable of breaking up to twenty feet while backing and ramming. Ships can run a combination of turbines or diesels in order to limit fuel consumption.

The oddest feature about the icebreaker's shiphandling characteristics is the three shafts with four-blade controllable-pitch propellers. The centerline and starboard shafts turn clockwise, while the port turns counterclockwise. In

Figure 13–15. The Coast Guard cutter *Healey* (WAGB 20) working in the ice. *U.S. Coast Guard Digital*

normal maneuvering this makes port turns easy and starboard turns more difficult. To increase maneuverability without significant headway, you can back the wing shafts while coming ahead on the centerline, allowing great rudder control. This is not recommended in the ice, as it places shafts at increased risk.

Maneuvering around piers can be a challenge, as this type of cutter is not responsive enough to adjust to the small changes required when mooring. Despite the controllable-pitch propellers, the 13,000-ton displacement makes the routine use of tugs a more prudent method of mooring and unmooring.

APPENDIX A

———◄o►———

INTRODUCTION TO THE
RULES OF THE ROAD

Every competent shiphandler requires a comprehensive knowledge of the Rules of the Road. Such a comprehensive understanding requires much more extensive treatment than can be provided here. This discussion should be considered only a preliminary introduction to the Rules, not an acceptable replacement for more detailed treatment as provided in some of the volumes listed in the bibliography.

International and Inland Rules

There are actually two sets of rules: the International Rules of the Road (also known as the COLREGS) and the Inland Rules. (Special rules also sometimes apply within designated waters and in the inland waters of other nations, but these exceptions are not covered here.) The International and Inland Rules are identical in most respects other than the meaning of certain sound signals, as discussed below. The discussion is in terms of the required sound signals, but in many cases it is prudent to parallel this by VHF radio.

Rules to Avoid Collision

From the shiphandler's point of view the most important rules are those specifying how ships are to maneuver to avoid collision. The Rules specify which ship is to maneuver to ensure safe passage (the *give-way* vessel) and which ship is required to maintain course and speed (the *stand-on* vessel.) There are two ways of determining this. One has to do with the characteristics of the vessels. Power-driven vessels are to keep clear of sailing vessels (except when the sailing vessel is overtaking the power-driven vessel), and vessels free to maneuver are to keep clear of those which are restricted in their ability to maneuver, for example, a large ship restricted to a narrow deep water channel, a dredge with suction down, ships connected for underway replenishment, and so on. The second way to determine which ship is to give way is through their relative positioning. The three cases of positioning identified in the rules are overtaking, meeting, and crossing situations. These are defined by the relative position of the two ships at the time they first sight each other visually.

Overtaking

Any vessel overtaking another is obligated to keep clear. A vessel is overtaking if she approaches the vessel ahead from more than 22.5 degrees (two points) abaft the beam. At night this would correspond to the situation in which the overtaken vessel's stern light is visible but the range, masthead, and sidelights are not. If doubt exists as to this angle, an overtaking situation will be assumed to exist. The vessel being overtaken, while legally the *stand-on* vessel, is permitted to conform to the channel.

The Inland Rules require an agreement by whistle signal between the two ships concerned before overtaking can take place. One short blast from the overtaking vessel proposes a passage to the starboard side of the overtaken vessel, and two short blasts propose a passage on the port side. If the overtaken vessel agrees, she returns an identical signal. If not, she must sound five or more short blasts. It is not permissible for the overtaken vessel to "cross signals" by answering a different signal than that proposed by the overtaking vessel. If the situation permits, the overtaking vessel may continue to try to reach an agreement with the overtaken vessel, but in the absence of mutual agreement it may not pass.

Under International Rules, the whistle signals represent actions, not proposals. One short blast means I am directing my course to starboard. Two short blasts mean I am directing my course to port. If the vessel being overtaken has doubts regarding the actions or intentions of the other vessel, it is obligated to sound five or more short blasts. To underline the rule, under both Inland and International Rules of the Road it is the obligation of the overtaking vessel to remain clear of the vessel being overtaken.

Meeting

A meeting situation exists when two vessels have their bows pointed at each other, or nearly so, and the range between the two is closing. The masts of the other ship would be nearly in line, and at night each ship would see both of the sidelights of the other. Under these circumstances, each vessel is to alter its course to starboard to leave the other on its port side as they pass.

Under Inland Rules, there is to be an exchange of whistle signals. Normally this will be an exchange of signals of one short blast, meaning I propose to pass you port to port. Again, cross signals are not permitted. If the second vessel does not agree with the proposal of the first, she should sound the danger signal of five or more short blasts. Although a starboard-to-starboard passage is permitted under the Inland Rules, under most circumstances passage should be port to port. Only if the ships are clearly set to pass safely starboard to starboard without altering course and agreement has been reached between the two vessels should a starboard-to-starboard passage be selected.

Under International Rules, the whistle signals are again tied to action rather than proposal. As under the Inland Rules, the majority of passages in meeting situations should be port to port. Under most circumstances, a visible alteration of course to starboard, accompanied by one short blast, will serve to communicate to the other vessel your intention to pass port to port. This action should be taken early to avoid possible misunderstanding. It is again desirable to parallel this with an understanding by VHF radio. In meeting situations, a port-to-port passage is almost always to be preferred.

Crossing

If it isn't an overtaking or meeting situation, then it must be crossing. The basic rule for a crossing situation is that the vessel that has the other on her right hand is the *give-way* vessel, and the one having the other on her left hand is the *stand-on* vessel. It is the obligation of the *stand-on* vessel to maintain course and speed and of the *give-way* vessel to maneuver as necessary to remain clear. This may be done by changing course or slowing to go under the stern of the *stand-on* vessel. Both International and Inland Rules state that the *give-way* vessel "shall, if the circumstances of the case admit, avoid crossing ahead of the other vessel" (International Rule 15, Inland Rule 15 [a]). Thus although not explicitly forbidden in the rules, it is rarely a good idea for the *give-way* vessel to speed up or come left in order to avoid the other.

Under Inland Rules, the *stand-on* vessel may sound one blast to indicate her intention of maintaining course and speed. The *give-way* vessel should respond with one blast to indicate her intention of coming right or slowing as necessary. In International Rules whistle signals would again indicate action being taken: one blast to indicate "I am coming right" or three blasts to indicate "my engines are backing." In either Inland or International Rules, if either ship fails to understand the movements or intentions of the other she should sound five or more short blasts.

At one time the Rules of the Road required that the *stand-on* vessel hold course and speed to the point of extremis, defined as the point at which collision can only be avoided by both vessels taking action. This could make it very uncomfortable for a ship required to stand on while the vessel required to give way was taking no action. The current rules provide more leeway. Rule 17(a)(1) states that the *stand-on* vessel "may, however, take action to avoid collision by her maneuver alone, as soon as it becomes apparent to her that the vessel required to keep out of the way is not taking appropriate action in compliance with these Rules." The rules go on to provide that if the *stand-on* vessel does elect to take avoiding action, she "shall, if the circumstances of the case permit, not alter course to port for a vessel on her own port side" (International and Inland Rule 17 [c]).

The General Prudential Rule

Not all circumstances fall neatly into the situations defined by the Rules. It is entirely possible, for example, to find yourself in a situation in which you are the *stand-on* vessel with regard to one contact, and the *give-way* vessel with regard to another. In such a situation it is impossible to comply with the rules for both. The Rules provide for this with the *General Prudential Rule*, also known as the Special Circumstances Rule. Both sets of rules state, "In construing and complying with these Rules due regard shall be had to all dangers of navigation and collision and to any special circumstances, including the limitations of the vessels involved, which may make a departure from these Rules necessary to avoid immediate danger" (International and Inland Rule 2 [b]). This provides the shiphandler with a certain amount of leeway in applying common sense to sort out complicated situations, but only when "necessary to avoid immediate danger."

The Rules of the Road, although having the force of law, are not intended to be unbendingly rigid. Interpreters of the Rules have recognized an underlying expectation of the practice of good seamanship. Not only is responsibility for complying with the rules spelled out, but they go on to say nothing shall exonerate "the neglect of any precaution which may be required by the ordinary practice of seamen, or by the special circumstances of the case" (International and Inland Rule 2 [a]). The competent shiphandler must know and comply with the Rules but use them with sound nautical judgment.

APPENDIX B

———◦———

PROPOSED STANDARDIZED TUG COMMANDS AND OPERATING PROCEDURES AS THEY APPLY TO ASSISTS AND ESCORT TUGS

Capt. Victor J. Schisler

(Basic Version Only)

(1) Reference tug commands to own ship. Specify the direction relative to the assisted vessel, i.e., "Guard pull easy to port."

(2) When asking for a push or pull, include the direction that the force is to be applied, port, starboard, forward or aft, e.g., "Pull easy to starboard," "Pull easy to starboard 45 degrees forward (aft)."

(3) The majority of these commands apply to conventional tugs as well as "tractor," "combi-tug," and "ASD" (azimuthing stern drive) type tugs.

(4) Always give tug name before giving commands. This will alert the tug captain that what she/he hears next will apply to her/his tug.

(5) Do not use given names of tug operators, as this may leave the ship's bridge team out of the information loop and there may be more than one "Bob" working tugs at that time.

(6) The tug operator should attempt to maintain the last angle requested by the pilot until changed by another command, e.g., the pilot may request that you "stop at 90" or "stop and drag on your line."

(7) Use power references of EASY (for 1/3), HALF (for 2/3) or FULL (for 100%).

(8) When the assist tugs are equipped with strain gauges, use line pull in tons if desired instead of "EASY," "HALF," and "FULL."

(9) When the tug has a linehandling winch, the towline to the stern of an assisted vessel should be at least as long as the tug. The amount of maneuvering space will determine the maximum length acceptable. This minimum length will allow the tug to work outside the propeller wash in most cases. The pilot may ask for more or less length, depending on the maneuvers planned.

(10) Conventional tugs with linehandling winches should use a long line to the stern of the ship and if requested shorten up to a "push-pull mode."

(11) The closer the towline is to the horizontal, the less tug weight is added to the towline on pulls.

(12) When tugs are utilized ahead of the ship, keep the following cautions in mind:

a) All tugs are limited in their ability to maneuver around the bow of a ship at speeds over five knots.

b) At large angles to the bow the tug is at risk of "getting in irons" or being "tripped."

c) The tug's ability to influence the ship's heading is limited as the pivot point moves forward.

d) The higher the ship's speed through the water, the more inline the tug must stay.

e) With a limited working angle at high speeds, the majority of the tug's line pull is causing the ship to accelerate.

(13) The "working end" of the tug is defined as the position that the working line comes from, i.e., the end that the linehandling winch is installed on.

BASIC TUG COMMAND	PURPOSE	MANEUVER	ADJUSTMENTS
1. PULL INLINE ASTERN (EASY, HALF, OR FULL POWER)	To control the speed of the assisted vessel.	Reverse the thrust of the tug and apply the towline pull inline with the keel of the assisted vessel.	Use steering thrust to maintain the alignment with the assisted vessel.
2. INDIRECT PORT (STARBOARD) (SOME ANGLE IN DEGREES)	To reduce the speed and to influence the heading of the assisted vessel. This maneuver is most effective above six (6) knots. However, it can be used by most tugs at any speed that does not cause deck edge immersion.	Use steering thrust to orient the towline to a relative angle of 45 degrees to the keel of the assisted vessel. Use the fore and aft thrust to achieve this. Orient the tug to the water flow in order to maximize the hydrodynamic lift of the hull. This is the "pure" indirect mode.	Initially, place your tug at 45 degrees (the pilot may request more or less angle) to the keel of the assisted vessel. Then, maximize the hydrodynamic lift of the tug's hull by adjusting the angle of your vessel to the flow of water. Do not allow the hull to "stall." Do not use any more power than is required to hold the requested angle.
3. BOW (STERN) TO STARBOARD (PORT) EASY, HALF, OR FULL	To control the speed and heading of the assisted vessel.	Pull off to the side requested at the power setting requested.	This allows the tug captain to determine the method.

BASIC TUG COMMAND	PURPOSE	MANEUVER	ADJUSTMENTS
4. PUSH BOW TO PORT (STARBOARD) EASY, HALF, OR FULL	To influence the direction of movement of the bow.	Apply thrust toward the assisted vessel by pushing on the bow.	Push at 90 degrees to the keel (not the hull plate you are looking at) of the assisted vessel unless another angle is requested.
5. PULL BOW TO STARBOARD (PORT) EASY, HALF, OR FULL	To influence the direction of movement of the bow.	Apply thrust away from the assisted vessel by pulling on the line.	Pull at 90 degrees to the keel (not the hull plate you are looking at) of the assisted vessel unless another angle is requested.
6. COME TO A PUSH-PULL MODE AND STOP	To shorten the reaction time of the tug when asked to push on the assisted vessel.	Shorten your towline, move the tug close to the assisted vessel, and be ready to push or pull as requested.	Do not confuse this with "push full." On high leads, be careful when adding the weight of your tug to the line on pulls away from the ship. When possible, slack out enough line to achieve an angle of less then 45 degrees above the horizontal.

APPENDIX C

———◦———

SAMPLE STANDING ORDERS
FOR AIR OPERATIONS

Reproduced with the permission of Vice Adm. James Stavridis, USN, from his *Watch Officer's Guide*, 14th ed. (Annapolis: Naval Institute Press, 2000), 332–37, 340–41.

Planeguard Operations

I will never criticize an OOD who maneuvers FISKE out of planeguard station and into open sea room because he or she is uncertain of the carrier's aspect, movement, or intentions.

1. *CV Operations.* Operations in close proximity to an aircraft carrier landing and launching aircraft require extraordinary vigilance and adherence to prudent seamanship to ensure the safety of the ship. When assigned as a planeguard, FISKE must be ready to recover a downed aviator or man overboard at a moment's notice. Nothing must be permitted to delay execution of this critical mission. Prior preparation is required, as well as frequent rehearsal of planned emergency actions.

2. When operating with CVs, stay out of a moving envelope 6,000 yards ahead, 4,000 yards abeam, and 2,000 yards astern unless directed to a station within this envelope by competent authority and I am on the bridge. Never turn toward a CV during maneuvers. If in doubt about a carrier's aspect or course during maneuvers, turn away to open range and call me immediately.

3. *Preparations.* The Officer of the Deck is fully responsible for making the ship ready to rapidly recover a downed aviator, and taking and maintaining

assigned stations. In preparing for planeguard operations the following guidelines, though not all-inclusive, should be considered:

a. Upon notification of impending planeguard operations, or two hours prior to the start of land/launch operations, inform me of your intentions and begin timely preparations.

b. Review the standard operating procedures for the specific carrier involved and note specific actions required by you or your watch team.

c. Muster the lifeboat crew and ensure they are instructed on specific procedures and your intentions with regard to type of pickup, maneuvering, etc. Personally instruct the Petty Officer in charge and Boat Officer on their duties. Ensure portable radios are ready and tested and sound-powered phones have been connected and checked. At night, provide and test necessary lighting.

d. Research the appropriate lighting measure. Unless otherwise directed by a specific carrier's SOP, lighting measure GREEN will normally be set during actual flight operations. This includes side lights set on dim, aircraft warning lights, a blue stern light, and no masthead lights. Ensure normal running lights are displayed on bright except when actually conducting flight operations, and especially when the carrier is reversing course or otherwise maneuvering. Follow the motions of the carrier in this regard by observing shifts between white and blue stern lights, but do not hesitate to turn normal running lights on bright if the carrier begins to maneuver unexpectedly or you otherwise are uncertain as to the carrier's intentions or your position. In such circumstances, inform me immediately.

e. Be forehanded in bringing up required communication circuits such as CCA, land launch, departure, marshall, the primary maneuvering circuit, and the designated Bridge-to-Bridge channel. Obtain radio checks as operations and the situation permit, but do not fill the airwaves with repeated call ups.

f. Know the station you are to be assigned, or query the carrier early enough to enable you to proceed to station so as to be in position at the appointed time or thirty minutes prior to sunset. Anticipate carrier maneuvers up- or downwind to ensure you are not grossly out of position and end up in a long tail chase.

4. *Stationing.* Normally carriers will assign a station 170 degrees relative, one thousand to two thousand yards astern. 170 degrees relative is typical.

a. Station limits in bearing remain 2 degrees. If in 170 degrees relative position, simply position yourself in line with angled deck line up lights and you should be well within limits.

b. Since the carrier can be expected to make frequent speed changes, station limits in range will normally be extended from two hundred yards short to five hundred yards long in range. Make speed changes boldly to remain with limits set by me. Remember, the carrier's reported and actual speed may be several knots different. Do not stop adjusting speed until the measured range rate is zero.

c. I expect you to stay on station. Continue to aggressively drive the ship toward the exact point station. Do not be an incrementalist. Maneuver boldly. Change course by 5 or 10 degrees or more, and speed by five or more knots until you see the relative motion required to get you to station. Watch bearing drift and range rate carefully. Keep checking and rechecking. Do not assume a specific course and speed for the carrier until you have verified her actual course over several minutes and can control your closing and opening rate at will.

d. The best estimate of the carrier's course and speed is the data reported over Link 11 tempered by your own observations and judgment. Ensure the carrier's PU is present on a surface contact ASTAB. Also ensure an ARPA track is initiated on the carrier. Note tactical signals and flashing light messages and obtain the best possible data from the DRT or maneuvering board, but do not rely on these aids alone. Watch the carrier's aspect, bearing drift, and range closely to obtain course and speed.

e. Ensure you anticipate carrier movements and react in a timely manner. If the carrier turns toward your side, turn away. If the carrier turns away from you, you can usually safely parallel the carrier's movements. Be alert for the unexpected.

5. *Maneuvering.* Expect the carrier to do the unexpected. Never assume she will turn or slow until you actually see a change in aspect, bearing drift, or range. Anticipate the carrier's next turn, especially during cyclic operations.

a. The objective is to remain astern, or off the quarter of a maneuvering carrier. Do not allow yourself to get ahead of the carrier's beam; take action to preclude the carrier turning inside of you when she reverses course. Remember, the carrier may turn greater than 180 degrees, or even 360 degrees, to reduce angle of heel or find a better wind.

b. If you are on the carrier's starboard quarter (e.g., 170 degrees relative, two thousand yards) and the carrier turns to port away from you, simply maintain your original course and follow around outside the carrier's wake. Increase speed markedly if necessary to close the turning point, and/or turn outboard to prevent the carrier from closing your projected course when coming all of the way around. Do not turn too early or slow. Watch the carrier's wake and follow around. At night, use the three-minute rule to calculate time to turn and start a stop watch when the carrier signals its rudder is over or when the turn is detected first by Bridge personnel.

c. If you are on the carrier's starboard quarter (e.g., 170 relative, two thousand yards) and the carrier turns to starboard toward you, immediately put on left standard rudder and come 40 degrees left of original course. Cross through the carrier's wake and then come back right, following the carrier through the turn just outside the carrier's wake as above. Remember, the OOD on the carrier will be looking for your starboard running light so he or she knows you have turned toward the carrier's wake and will be well clear. Show your sidelight smartly.

d. If the carrier increases speed, as determined by an opening range or tactical signal, increase speed markedly so you will not get left behind. It is always easier to slow than regain lost ground. However, do not close inside your inner station limits without my permission.

e. If the carrier slows, as detected by a decreasing range or tactical signal, decrease speed slowly as we will tend to slow more quickly than the much larger carrier. If range decreases noticeably, slow five or ten knots to regain control of your range rate. If you reach the inner edge of your station, slow to five knots or bare steerageway until range begins to open again, then quickly resume ordered course. Do not hesitate to stop, sheer out, or back down to preclude closing dangerously close. Keep me fully informed.

f. Ensure regular running lights are displayed on bright whenever the carrier and/or we are maneuvering.

g. Do not hesitate to use the radio to resolve a developing emergency situation. If you take such action, immediately inform me and:

1. Use plain language and ships' names vice tactical signals and call signs, e.g., "AMERICA this is FISKE, my rudder is right. My speed is 12. I do not understand your intentions."

2. Always give your course, speed, and intentions.

3. Use any circuit on which you have good communications with the carrier's bridge. Usually this will be the tactical circuit in use or designated Bridge-to-Bridge channel.

4. Use the whistle to signal your actions.

h. Do not hesitate to take action to resolve an emergency situation. Inform me immediately, but do not let this reporting requirement detract from your primary responsibility for ship safety. If circumstances warrant, simply direct the Boatswain's Mate to pass "Captain to the Bridge" on the 1MC or sound the collision alarm.

i. If the situation warrants, or the carrier's maneuvers are unclear, turn away smartly, increase speed, and place the carrier astern to open your range. In all situations you should have an escape course in mind to use if the situation becomes unclear.

6. *Man Overboard.* In the event of an aircraft crash or man overboard, the objective is to recover the person as quickly as possible.

a. If a helo is assigned planeguard, it will normally be the primary recovery vehicle. However, be alert, especially at night, to assume this role if the helo becomes disoriented or two or more people are positioned in the water a good ways apart from each other.

b. Close the area to within about five hundred yards and stand by to assist. Remain well clear of the area upwind of the man if a helo is making the recovery.

c. Primary concern is to mark the area with a light/ring buoy and/or smoke float. If need be, order the carrier or helo to drop a marker.

d. Rehearse in your mind the myriad of specific procedures to take in various scenarios. Discuss the merits of a shipboard or boat recovery with me and get my concurrence with your intentions.

e. Use the best radio circuit to coordinate initial operations. Usually this will be the CCA, departure, or Land Launch in use. Emergency/SAR forces will not shift frequencies. Instead, forces involved in continuing operations should shift to an alternate frequency. Use plain language and any other means to facilitate a timely rescue.

f. At night, immediately break out lights and begin preparations for a lengthy search.

g. Ensure the best possible location of the crash or person in the water is plotted on the DRT and logged in the deck log.

Helicopter Operations

1. *General.* Helicopter operations shall be conducted in accordance with FISKEINST 3710.1 series (AVIATION STANDARD OPERATING PROCEDURES).

a. Anticipate the setting of Flight Quarters well in advance to permit thorough preparations and safety checks. Be aware of alert conditions established. Advise me when ready to launch, recover, or hover an aircraft and I will give "Green Deck."

b. Ensure proper communications are established among the Bridge, CIC, Helo Control Station, the crash detail, and the helo.

c. Verify the helo pilot is in possession of the latest navigational data.

d. Position the ship for desired wind conditions prior to engaging rotors. Do not request a green deck unless you are within the NWP-42 wind and roll/pitch envelopes for the type helo you are receiving. If you cannot obtain satisfactory winds and roll/pitch, ensure you apprise me.

e. Control the following specific actions from the Bridge, after obtaining my permission:

1. Green Deck: Helo operations authorized.
2. Amber Deck: Engaging and Disengaging rotors authorized.
3. Red Deck: Helo operations not authorized.

2. *Emergencies.* In the case of an emergency involving a helicopter under our control, set emergency Flight Quarters, close the helicopter's position at maximum speed (including coming to Full Power), determine the helicopter's problem, determine if we are the best platform to recover the helicopter and, if not, immediately contact the platform which may be, and notify our immediate Operational Commander. In the case of an emergency recovery the OOD may give a green deck. Should a helicopter not under our control declare an emergency and FISKE's TAO or OOD determines we are in a position to provide an emergency recovery, immediately contact the helicopter's controlling unit and offer FISKE's assistance. If appropriate to the circumstances (i.e., we are in the best position to provide assistance), simultaneously with offering our assistance to the helicopter's control unit, close the helicopter's position at maximum speed and set emergency Flight Quarters. In the case of such a recovery, the OOD may give a green deck.

3. *Wind and Sea Considerations.* NWP-42, Shipboard Helicopter Operating Procedures, provides guidance, operational procedures, and training requirements for the shipboard employment of helicopters. The Officer of the Deck shall be thoroughly familiar with all the requirements of this publication. In addition he or she shall:

a. Ensure the helicopter check-off list located in the Bridge OOD folder is completed prior to helicopter operations.

b. Maneuver the ship only during a "Red Deck" condition to obtain optimum wind, pitch, and roll conditions. The OOD must be cognizant of the tactical situation and maneuver the ship for minimum disruption of the formation or speed of advance while obtaining true wind from forward of the beam for all launch and or recovery operations.

NOTES

—◄◦►—

Introduction

1. *IMO Standard Maritime Communication Phrases* (London: International Maritime Organization, 2002), iii.
2. In some relatively rare circumstances, responsibilities may be divided as, for example, giving one officer control of the engines and another control of the rudder while alongside during an underway replenishment. The rationale for doing this is that there is little interaction between course and speed while alongside. Most experienced shiphandlers disapprove of the practice. A more common mode of splitting the conn is to have a second person controlling the tug or tugs. In this case, the officer controlling the ship's engines and rudders generally has the conn and tells a second officer what he wants done with the tugs, without giving specific orders.
3. Although it is common now to have a "helm safety officer" assigned during special sea detail to ensure that orders are executed as given, this provides little protection against a slip of the tongue by the conning officer.

2. Forces on the Ship

1. This is in part because when going astern the pivot point moves toward the stern, reducing the lever arm exercised by the rudder. Movement of the pivot point is discussed later in this chapter.

3. Standard Commands

1. Capt. James Stavridis, USN, *Watch Officer's Guide*, 14th ed. (Annapolis: Naval Institute Press, 2000).

4. Getting Under Way

1. Definitions are taken from Capt. John V. Noel Jr. and Capt. Edward L. Beach, *Naval Terms Dictionary*, 4th ed. (Annapolis: Naval Institute Press, 1978).

6. Ground Tackle

1. The Mediterranean or "Med" Moor is a method of mooring a ship using two anchors to secure the bow, with a stern line to a pier.

7. Transiting the Channel

1. Daniel H. MacElrevey, *Shiphandling for the Mariner* (Centreville, Md.: Cornell Maritime Press, 1983), 9.
2. Lt. Angus N. P. Essenhigh, RN, and Cdr. Michael T. Franken, USN, "Handling the *Arleigh Burkes*—Part Three," U.S. Naval Institute *Proceedings* 128, no. 6 (June 2002): 87.
3. See Inland Rules of the Road, Rule 34.
4. International Rules of the Road, Rule 10 (a).
5. Ibid., Rule 10 (b) (i.).
6. Ibid., Rule 10 (b) (ii).
7. Ibid., Rule 10 (b) (iii).
8. Ibid., Rule 10 (b).
9. Ibid., Rule 10 (h).
10. Ibid., Rule 10.
11. This is sometimes described as "pumping the rudder."
12. International Rules of the Road, Rule 9 (e) and 34 (c).
13. IMO Convention of Standards of Training Certification and Watchkeeping (STCW) A-VIII/2, para. 3.2.
14. COMNAVAIRFOR/COMNAVSURFORINST 3530.4, 26 February 2002, Appendix B.
15. Ibid., 2-10 and 4-9.
16. Ibid., Appendix B.

8. Tugs and Pilots

1. The *Fletcher* class had 20.5 shaft horsepower per ton, the *Sumner* class 18.6 shaft horsepower per ton, and the *Gearing* class 17.3 shaft horse power per ton at full load displacement. The *Arleigh Burke* class, depending on flight series, range from 10.8 to 12 shaft horsepower per ton.
2. COMNAVAIRFORINST/COMNAVSURFORINST 3530.4, 26 February 2002.
3. Ibid., 2-1.
4. Ibid., 2-2.
5. Malcolm C. Armstrong, *Pilot Ladder Safety* (Woollahra, NSW, Australia: International Maritime Press, 1979), 8–15.
6. A convention that you may encounter, but that is by no means universal, is for the tug to respond with one toot to all orders other than backing, and with two toots to backing orders.
7. Remember that the rudder steers by moving the stern in the direction opposite to the intended turn. If a tow line prevents the movement of the tug's stern, it takes away its ability to steer.

9. Underway Replenishment

1. Another way of getting a good handle on the actual speed being made good by the replenishing ship is to ask for her speed over the ground reading on the GPS.
2. This can be confirmed trigonometrically: The sine of 1 degree is .017, which is the same as 1/60.
3. A handy way to use the radian rule is to remember the constant 3,000. Dividing 3,000 by the range to the replenishment ship gives you the angle for a 150-foot separation. For example, 3,000 divided by a range of one thousand yards gives a 3-degree angle. Different constants can be used for other separations. If you are on a large hull and want a 200-foot separation, your constant will be 4,000.
4. Ens. John Payne, "Reducing the Risk of Unreps," *Fathom* (publication of the Naval Safety Center), July–September 2000.

10. Shiphandling in Emergencies

1. MacElrevey, *Shiphandling for the Mariner*, 175.
2. Capt. R. S. Crenshaw Jr., USN, *Naval Shiphandling*, 2nd ed. (Annapolis: United States Naval Institute, 1960), 144.

3. William J. Kotsch and Richard Henderson, *Heavy Weather Guide*, 2nd ed. (Annapolis: Naval Institute Press, 1984), 311.

11. Air Operations

1. USS *Enterprise* instruction 3505.1G, p. III-C-2.

12. Tactical Maneuvering

1. The text of the treaty may be found at http://dosfan.lib.uic.edu/acda/treaties/sea1.htm.
2. Two other useful rules of thumb are the three-minute rule and the one-minute rule. The three-minute rule is that each knot of speed equals one hundred yards in three minutes. Thus a ship traveling at twenty knots will cover two thousand yards (one nautical mile) in three minutes. The one-minute rule tells us that each knot of speed equals one hundred feet in one minute.

13. Special Ship Characteristics

1. Cdr. Terry D. Mosher, USN, "New Twist for the *Arleigh Burke*s," U.S. Naval Institute *Proceedings* 129, no. 9 (September 2003): 84–85.
2. Essenhigh and Franken, "Handling the *Arleigh Burke*s—Part Three," 87.
3. Cdr. James Stavridis, USN, "Handling the *Arleigh Burke*s," U.S. Naval Institute *Proceedings* 120, no. 10 (October 1994): 67.
4. Figures 13-6 to 13-9 provided by U.S. Navy Submarine School, New London, Connecticut.

BIBLIOGRAPHY

——◄◦►——

Becker, Cdr. John J. USN. "Special: The FFGs." Pts. 1 and 2. U.S. Naval Institute *Proceedings* 116, no. 1 (January 1990): 109–13, and 116, no.2 (February 1990): 99–106.

Crenshaw, Capt. R. S., Jr., USN. *Naval Shiphandling*. 2nd ed. Annapolis: United States Naval Institute, 1960.

Cutler, Thomas J. *The Bluejacket's Manual*. 22nd ed. Annapolis: Naval Institute Press, 1998.

Davis, Capt. Theodore F., USN (Ret.). "Shiphandling and Piloting Training *Ohio* Style." U.S. Naval Institute *Proceedings* 108, no. 2 (February 1982): 101–2.

Essenhigh, Lt. Angus N. P., RN, and Cdr. Michael T. Franken, USN. "Handling the *Arleigh Burkes*—Part Three." U.S. Naval Institute *Proceedings* 128, no. 6 (June 2002): 86–88.

Eyer, Cdr. K. S. J., USN. "Driving the Aegis Cruiser." U.S. Naval Institute *Proceedings* 128, no. 12 (December 2002): 70–72.

Faller, Cdr. Craig, USN. "Shiphandling Training? Ask Your JOS." U.S. Naval Institute *Proceedings* 129, no. 3 (March 2003): 104–5.

Hooyer, Henry H. *Behavior and Handling of Ships*. Centreville, Md.: Cornell Maritime Press, 1983.

International Maritime Organization. *IMO Standard Marine Communication Phrases*. London: International Maritime Organization, 2002.

Kotsch, William J. *Weather for the Mariner*. 3rd ed. Annapolis: Naval Institute Press, 1983.

Kotsch, William J., and Richard Henderson. *Heavy Weather Guide*. 2nd ed. Annapolis: Naval Institute Press, 1984.

Llana, Christopher B., and George P. Wisneskey. *Handbook of the Nautical Rules of the Road.* 2nd ed. Annapolis: Naval Institute Press, 1991.

MacElrevey, Daniel H. *Shiphandling for the Mariner.* Centreville, Md.: Cornell Maritime Press, 1983.

Maskell, Lt. Cdr. Robert E., USN. "Handling a *Spruance*-Class Destroyer: An Update." U.S. Naval Institute *Proceedings* 108, no. 4 (April 1982): 115–16.

Moran, Cdr. Gene, USN, and Lt. (jg) Amy Morrison, USN. "Handling the *Arleigh Burke*s—Part Two." U.S. Naval Institute *Proceedings* 126, no. 7 (July 2000): 85–86.

Mosher, Cdr. Terry D., USN. "New Twist for *Arleigh Burke*s." U.S. Naval Institute *Proceedings* 129, no. 9 (2003): 84–85.

Noel, Capt. John V., Jr., USN (Ret.). *Knight's Modern Seamanship.* 17th ed. New York: Van Nostrand Reinhold, 1984.

Noel, Capt. John V., Jr., USN (Ret.), and Capt. Edward L. Beach, USN (Ret.). *Naval Terms Dictionary.* 4th ed. Annapolis: Naval Institute Press, 1978.

Reid, George H. *Shiphandling with Tugs.* Centreville, Md.: Cornell Maritime Press, 1986.

Robinson, Lt. Scott A., USN. "Handling a Battleship." U.S. Naval Institute *Proceedings* 114, no. 4 (April 1988): 110–13.

Smith, Capt. Richard A., RN (Ret.). *Farwell's Rules of the Nautical Road.* 7th ed. Annapolis: Naval Institute Press, 1994.

Stavridis, Lt. (jg) Jim, USN. "Handling a *Spruance*-Class Destroyer." U.S. Naval Institute *Proceedings* 105, no. 10 (October 1979): 124–26.

Stavridis, Lt. Cdr. James, USN. "Handling a *Ticonderoga*." U.S. Naval Institute *Proceedings* 113, no. 1 (January 1987): 107–9.

———. Cdr., USN. "Handling the *Arleigh Burke*s." U.S. Naval Institute *Proceedings* 120, no. 10 (October 1994): 66–68.

———. Capt., USN. *Watch Officer's Guide.* 14th ed. Annapolis: Naval Institute Press, 2000.

Swinger, Cdr. Alan W., USN. "Getting a Handle on FFG-7 Shiphandling." U.S. Naval Institute *Proceedings* 108, no. 8 (August 1982): 106–8.

U.S. Coast Guard. Commandant Instruction M16672.2D, 25 March 1999. *Navigation Rules, International—Inland.* Washington, D.C.: U.S. Department of Transportation.

INDEX

◄○►

Page numbers in italics refer to photographs and illustrations.

ABOUT THE AUTHOR

———◄o►———

Jim Barber was commissioned into the Navy from the NROTC Regular Program at the University of Southern California, served nearly thirty years on active duty, and retired as a captain. During that time he was special sea detail OOD on an aircraft carrier, senior watch officer on a destroyer, and executive officer of a destroyer, and he commanded a destroyer escort, a guided missile frigate, and a guided missile cruiser. In between sea tours he had a variety of assignments, all of which were interesting, but none were as much fun as shiphandling. Following retirement from active duty he served as a convoy commodore and as CEO and publisher of the U.S. Naval Institute from 1984 until 1999. He enjoys teaching beginning sailing in Annapolis.

Captain Barber earned a PhD at Stanford University and has taught at Vanderbilt, the Naval War College, and George Washington University. He currently teaches for the Naval Postgraduate School in the U.S. Naval Academy's Leadership Education and Development Program. His awards include the Navy League's Alfred Thayer Mahan Award, the Defense Superior Service Medal, the Legion of Merit, the Bronze Star with combat "V," the Meritorious Service Medal with gold star, and six awards of the Vietnam Service Medal. In 1999 he was presented with the U.S. Coast Guard's Meritorious Public Service Award, and in 2000, with the Navy's highest civilian award, the Distinguished Public Service Award. This is his third book.

The Naval Institute Press is the book-publishing arm of the U.S. Naval Institute, a private, nonprofit, membership society for sea service professionals and others who share an interest in naval and maritime affairs. Established in 1873 at the U.S. Naval Academy in Annapolis, Maryland, where its offices remain today, the Naval Institute has members worldwide.

Members of the Naval Institute support the education programs of the society and receive the influential monthly magazine *Proceedings* and discounts on fine nautical prints and on ship and aircraft photos. They also have access to the transcripts of the Institute's Oral History Program and get discounted admission to any of the Institute-sponsored seminars offered around the country.

The Naval Institute also publishes *Naval History* magazine. This colorful bimonthly is filled with entertaining and thought-provoking articles, first-person reminiscences, and dramatic art and photography. Members receive a discount on *Naval History* subscriptions.

The Naval Institute's book-publishing program, begun in 1898 with basic guides to naval practices, has broadened its scope to include books of more general interest. Now the Naval Institute Press publishes about one hundred titles each year, ranging from how-to books on boating and navigation to battle histories, biographies, ship and aircraft guides, and novels. Institute members receive significant discounts on the Press's more than eight hundred books in print.

Full-time students are eligible for special half-price membership rates. Life memberships are also available.

For a free catalog describing Naval Institute Press books currently available, and for further information about subscribing to *Naval History* magazine or about joining the U.S. Naval Institute, please write to:

Membership Department
U.S. Naval Institute
291 Wood Road
Annapolis, MD 21402-5034
Telephone: (800) 233-8764
Fax: (410) 269-7940
Web address: www.navalinstitute.org